Still Water

www.penguin.co.uk

Still Water

The Deep Life of the Pond

John Lewis-Stempel

doubleday

TRANSWORLD PUBLISHERS
61–63 Uxbridge Road, London W5 5SA
www.penguin.co.uk

Transworld is part of the Penguin Random House group of companies
whose addresses can be found at global.penguinrandomhouse.com

First published in Great Britain in 2019 by Doubleday
an imprint of Transworld Publishers

A CIP catalogue record for this book is
available from the British Library.

ISBN 9780857524577

Typeset in 12.75/17 pt Goudy Oldstyle Std by Jouve (UK), Milton Keynes
Printed and bound in Great Britain by Clays Ltd, Elcograf S.p.A.

Penguin Random House is committed to a sustainable
future for our business, our readers and our planet. This book
is made from Forest Stewardship Council® certified paper.

1 3 5 7 9 10 8 6 4 2

Not a fish can leap or an insect fall on the pond but it is thus reported in circling dimples, in lines of beauty, as it were the constant welling up of its fountain, the gentle pulsing of its life, the heaving of its breast.

Henry David Thoreau, *Walden*

Contents

Prologue

I swim with slow strokes, enough to keep me moving forward, not sinking.

This is the farm pond, seen anew: the frog's eye view. Breast-stroking, my legs scissoring behind me, I feel amphibian.

I make no sound as I cruise in a circle, a twenty-five-metre ripple-ring.

The water is cold silk, closing around my back as I finish each stroke.

In modern argot I am 'wildswimming'; as a child in the 1970s, our gang's trips to the River Lugg at Mordiford were merely 'swimming'. Everything that was once ordinary has to be übered.

When I came up to the pond half an hour ago it was as flat as a silver coin; a breeze has begun to ruffle the water. On the bank, tall swords of yellow iris, already in flower, sway tipsily.

But it is dreamy warm, and the air is lazy with the drone of hoverflies in those daffodilly iris.

You might think a swim like this, in a pond with the midges as thick as mesh, silly. But I have an affinity with ponds, I feel the instinct strongly to visit every one, and pootle about their edge. And then it occurred to me this fine June morning, that perhaps the real way to enjoy the small, still waters is from within. From the centre looking out, not the outside looking in. To become pond life.

A damselfly flickers around; a red-hot darning needle probing the sedge.

The moorhen momentarily peers out from under the overhanging sallow, then withdraws to her shadow-cave to continue her disconsolate clucking unseen. The great white frog, me, is still there.

Lapping water disturbs a mat of fallen iris stems; a real frog plops off, down into the plant's roots, dark and thick as old ship's rope.

The pond is brim-full, from the winter's rain and the spring that feeds into its base. The water is at its clearest now; come August, summer sun will have reduced the level, and the pond will have its familiar stink of wet dog, rotting veg.

My eyes two inches above the surface of the water, I swim another circle of this closed world, with its circular bright green rim of iris, sedge and waterweeds; wild celery, water crowfoot, brooklime, water forget-me-not, water dock, water starwort, water plantain, water speedwell.

A breath of watermint comes over the surface of the pond, from leaves crushed by the cautious fox when he came down to lap the night water. Watermint is concealment for Mr Tod, both for its physical structure and its perfume. The herb, in medieval times, was strewn on the floors of houses to disguise smell.

Alder trees top and tail the pool, sheltering it from the north wind and the south sun and making it attractive to mallard. There is a clump of rabbity bramble too. Wild duck dislike open water inland. The resident female mallard, made flightless by motherly love, paddles through the edge scum and algae, which break apart to leave a clear path for her three bobby ducklings. Duck and duckettes exit by the mud slipway, where the cattle come down to drink in August; the birds disappear away into the buttercups.

Today, the Friesian cows stand on the steep adjacent hill, as though stacked on each other's backs in an acrobats' pyramid. The hill is the one topographical object visible above the herbage of the pond. There is only the hill, the sky, the pond and me.

A newt hangs in the water, regards me, then zips

off. A pied wagtail totters back and forth on the cows' access ramp, absorbed in her insecty business. A darter hitches a lift on my nose; a member of the dragonfly family, the darter is among the most ancient species on the planet, 300 million years old. Dragonflies endow a pond with timelessness.

I am the frog-man, and in this guise the majority of the animals and birds feel no threat from me. It is the same easy closeness one gets to wildlife on the back of an equine, when one is a horse-man.

As I bring my arms together for another pull through the water, a backswimmer (*Notonecta glauca*) rises in front of my face, before diving down again, a beetly mercury oblong. The backswimmer holds air bubbles on its chest; nature's version of oxygen cylinders.

In this pond of verdancy and forgetfulness, I am not entirely at ease. The backswimmer, for all its air-carrying cleverness, is a 1.5cm carnivorous monster which stabs its victim with its feeding beak (rostrum), before injecting a toxic juice that paralyses and liquefies the innards of said victim. Tadpoles are a staple food.

I was once bitten by a backswimmer; it's as bad as being stung by a bee.

Down somewhere in the depths slithe the carp,

great grey submarines. I fear the touch of their alien skin.

In the pond there are wondrous, multitudinous life forms. And curious cruelties.

Swimming in the pond, I find, is part pleasure, part frisson; part surface-seen, part submerged-mystery. I am half in my element in the pond; after all, about 50 per cent of the human body is made up of H_2O.

It's semi-safe, the pond. A womby sac compared to a lake.

I suppose some would do this swim naked, properly amphibian. I, however, am an Englishman, thus am wearing floral swimming trunks.

As I make one last circle, I hear a distinct vibrato from under the water. The male lesser water boatman (*Corixa punctata*) 'stridulates' by rubbing its front legs against a ridge on its head, a similar 'chirping' technique to that of terranean grasshoppers.

It is nature's water music.

Introduction

They were always there, so I took them for granted. My one excuse is that their humility was their own enemy. They were in the back of the landscape, the corner of the mind, unlike the foreground (and money-making) meadow I 'hayed', the wood I ran pigs in, the arable field I harvested, the hillside on which the sheep clung. Such is the blind preoccupation of the farming life.

I was brought up short by George Orwell's *Coming Up for Air*, his 1939 novel in which suburban insurance salesman George Bowling tries to recapture his youth in Lower Binfield, beside the tranquil fishponds of his childhood.

In *Coming Up for Air* ponds are the defining image of childhood. Of England too.

Looking back to his boyhood fishing the cowy, carpy ponds at Mill Farm and Binfield House, Bowling (a thinly fictionalized Orwell) sees clearly everything that was right about a childhood in nature, and every-thing that is wrong about the modern adulthood alienated from nature:

> I wondered why it is that we're all such bloody fools. Why don't people, instead of the idiocies they do spend their time on, just walk round LOOKING at things? That pool, for instance – all the stuff that's in it. Newts, water-snails, water-beetles, caddis-flies, leeches, and God knows how many other things that you can only see with a microscope. The mystery of their lives, down there under water.

As I read *Coming Up for Air* – a significantly aqua-tic title – I revisited in my mind the ponds of my own childhood: the lily-and-goldfish pond my father made, complete with flowing stream down the rockery (and into which my cousin Benji fell, stealing the limelight at my fifth birthday party), the ponds at the Quarry, the cow pond in front of my grandparents' house, the duck pond by the Castle Green in Hereford . . . And all the ponds I had ever known cascaded into my memory: Kilpeck moat, the dew pond above our cottage at Abbeydore, Cwm Farm pond, the dew pond by the Iron Age farmstead on Garway Common, the pond by the bandstand on Clapham Common, the mires on the top of the Black Mountains, the pond I made with the mini-digger on our farm in the Black Mountains, the kitchen sink, the wedding-present terracotta birdbath from my sister-in-law . . .

That terracotta birdbath poses a crucial question: what is a pond? There are official explications: according to the Freshwater Habitats Trust, previously Pond Conservation:

A pond is a small area of shallow (generally <2m), still, freshwater less than 100m across. Due to their small size and a lack of wind mixing, ponds are prone to chemical stratification of the water, resulting in high oxygen and pH at the water surface and low oxygen and pH towards the pond bed. In larger shallow water bodies this chemical separation of the water does not usually occur. Ponds can be permanent or temporary, lasting for only a few months in a given year.

So a puddle can be a pond, as can a ditch ('a linear pond'). And a large birdbath. 'Limnologists', those who study freshwater habitats, would add that ponds tend to have 'high aquatic biodiversity'.

Ponds are both naturally occurring – such as the 'pingos' of East Anglia, scoops sculpted by the retreating glaciers – and artificial. The types of human-made pond are many: there are fishponds, garden ponds, stock ponds, filled-in-with-water marl pits, moats, saw pits in woods, charcoal pits in forests. Indeed, etymologically 'pond' seems not to appear until the Middle Ages, and then in the specific sense of dammed water. The staple Anglo-Saxon word for a small expanse of water was *mere*, and occurs in place names galore, as suffix and prefix: Buttermere, Tangmere, Merton.

In the British countryside, man's pond predominates. According to the landscape historian Oliver Rackham, in unpopulous hill zones of England there are usually fewer than two ponds per square mile. In agricultural areas of lowland England, there are twice as many. The Victorian century was the golden time of the British pond. The number of ponds over twenty feet across in England and Wales in 1880 was about 800,000, or fourteen ponds per square mile, and in East Anglia it was thirty ponds per square mile.

Of course, it is not enough to define natural phenomena in abstract technical terms. It is never enough. A pond is a place of still water; the surface of a pond might be ruffled, or corrugated, but it will never be wavy. A pond is a place where fallen leaves will circle like toy yachts because of wind rather than current.

I, at least, define ponds by their birdlife. The moorhen is a pond bird; its relative the coot is a lake bird. The moorhen, suspicious of large bodies of water, when threatened heads for the edge cover; the coot lurches for open water.

In the life cycle of the human, the pond has meaning at particular stages. It is especially a place of childhood, as Orwell noted, and as Edward Thomas also knew. 'A pond needs nothing else except boys like us to make the best of it,' Thomas wrote in his autobiographical novel, *The Happy-Go-Lucky Morgans*.

After childhood one walks away from the pond, venturing over dramatic mountains, along dangerous rivers. One returns to the pond as a parent – to feed the ducks, to go pond 'dipping' with the eternal aids of a net and jam jar. This is pond-loving by proxy.

What we made, we taketh away. We have 'lost' around half a million ponds in the last century. Although there are still around 470,000 ponds in the British countryside, 80 per cent are polluted and degraded by lack of maintenance. (In Scotland, only 10 per cent are polluted, and it is estimated that half are in good condition.) They are polluted by nutrients, heavy metals, sediments, biocides, excess fish, excess waterfowl. Two thirds of ponds are polluted by just two agricultural chemicals – nitrogen and phosphorus, key ingredients in artificial fertilizer. Nutrient enrichment from fertilizer, contained in field run-off, results in deadly dominance by planktonic algae (making the water green) and filamentous algae, or 'blanketweed'. Our chemical war on ponds was already well underway when Orwell wrote *Coming Up for Air*:

> When I was a kid every pond and stream had fish in it. Now all the ponds are drained, and when the streams aren't poisoned with chemicals from factories they're full of rusty tins and motor-bike tyres.

Dead water. The deadness of dead water can be assessed by chemical sampling. Alternatively, one can look at the pond and see the presence or absence of life. Ponds support two thirds of all freshwater species including the common frog, common toad, teal, common great diving beetle, pond olive mayfly, blue-tailed damselfly, broad-leaved pondweed, great crested newt, pillwort, and medicinal leech. (In Britain, more than a hundred priority species are associated with ponds.) Go dip your net in a pond, and it will scoop anything from water scorpions to caddisfly larvae. A dip with a net in a pond is the original lucky dip.

Ponds, by and large, need human care. They are slow suicides. By the procession of 'succession' they fill with soil, leaf litter, detritus, to become scrub.

Humans assist that suicide with ways other than chemicals. When Orwell's Georgie Bowling returns to Lower Binfield, he finds that:

> The Mill Farm had vanished, the cow-pond where I caught my first fish had been drained and filled up and built over, so that I couldn't even say exactly where it used to stand.

Things are no better when Bowling goes to Binfield House. One pond has been turned into a manicured lake for residents of the new 'posh' estate – with its sham-Tudor architecture – to sail model boats, while the other is a rubbish-filled dump:

> They'd drained the water off. It made a great round hole, like an enormous well, twenty or thirty feet deep. Already it was half full of tin cans.
>
> I stood looking at the tin cans . . .
>
> God rot them and bust them! Say what you like – call it silly, childish, anything – but doesn't it make you puke sometimes to see what they're doing to England . . .

Coming Up for Air is a lament for the loss of child-hood, and for a wildlife-rich England that was already passing into memory by the 1930s.

It is the ultimate degrading folly, of course, to degrade the pond. We came from the primordial pond.

Call it late, but after reading *Coming Up for Air* two years ago I thought I should give something back to ponds, by writing about them. By putting them in full view I could, perhaps, make good my lifetime of careless assumption. Maybe I had not treated ponds reverently, but they had always been on my mind.

I rarely leave the Herefordshire farm. My books are about the place I work. With *Still Water* I intended to visit ponds up and down Britain, and write a watery travelogue. Well, I have indeed managed to get off the premises – even as far as France, in a series of 'study leaves' on organic, sustainable agriculture – but the more I looked at ponds, the more I realized I was taking the wrong route.

The necessary journey was to go down, down into the nature of the pond. Depth, not breadth. Submerge myself in the pond's seasonal cycle, dredge its history.

I listened to Orwell:

You could spend a lifetime watching them, ten lifetimes, and still you wouldn't have got to the end even of that one pool. And all the while the sort of feeling of wonder, the peculiar flame inside you. It's the only thing worth having . . .

So, this is my book of pond praise and plea.

THE POND IN WINTER

The Farm Pond, Herefordshire
8 NOVEMBER: Mild, cracked-paint sky, hateful in its
deadness. Across the meadow, the grass gnarled to the
bone by the sheep, the pond, usually the focal point,
now something secondhand, a dirty forgotten mirror.
In among the broken rushes, the moorhen walks
gingerly, then sees me. She flees, flailing across the
water, scarring it with her legs, so it vibrates with brief
life. There's an ambiguity in the moorhen's flight; does
the water help or hinder her? The trailing feet in the
water are symbolic of course: the moorhen is bound to
the pond. Along with the mallard, it is the pre-eminent
pond bird. The word 'moor' here is an old sense
meaning of marsh; the species is not usually found on
moorland. Moorhens are territorial creatures, often
faithful to a particular pond.

On the pond bottom: sunken leaves and sticks,
the decaying leftovers of summer. The ripples from the
moorhen settle; then there is nothing.

We inherited this pond in a field; on receipt, it
was half full-up with mud, meaning it was, in terms of
'biodiversity', firing on two out of four cylinders. Last
autumn I dredged one end with a Kubota mini-digger,
down to the clay, dumping the stinking sludge on the
bank. A troupe of bleary, bewildered frogs stumbled
out; they had already gone into hibernation, stalling

19

their amphibian metabolism, so they were neither quite dead, nor quite alive; they can survive underwater in stasis because they are able to breathe through their skin as well as through their lungs.

The frogs forgave the interruption, and in numbed, slow-motion hops re-entered the pond. The larvae of insects, the beetles, the nameless crawlies were less fortunate. For three days the buzzard and the pied wagtails picked over the heaving grey waste until they were bloated, the mud stuck to the gorging buzzard's breast like a bib.

Early the next morning, in the murk, checking the sheep, I see the heron looking into the pond, bemused, wondering where all the teeming amphibians of yesterday have gone. Down deep, Big Bird, down into the mud, and the forgetful sleep of hibernation, and the mysteries of metamorphosis.

The Mill Pond, Argenton, Western France
12 NOVEMBER: Drive from Caen to Argenton in the Deux-Sèvres department, through the long night. The land between the port and village is dark and very old. The only light in the villages from the *boulangerie*; the

plane trees along the road as regular as railings in the car headlights.

Arrive Argenton 8.30am (French time; an hour ahead). Mist rising through sunlight, the mill pond a hot molten pool; only a noddy-headed moorhen travelling the surface proves the blazing visual falsehood, together with my wellingtoned feet dangling in the water. The mill pond, with its pollarded ash, is set against an impossibly romantic ruined mill, with dovecot, creeping ivy, ferns, the spirit of Byron. The water from the Argenton river spouts from a pipe in the mill race, snails over flat stones, through reeds, to trickle into the pond with the sound of tap water into a bath. I watch the bubbles to see how far they will go on their voyage before dying, before they burst.

Grass and docks on the smooth sandy bank under the ruin; most of the rest of the pond edge is perpendicular, two feet high, straight down, with vole holes peering through lank, green fringes of grass; at the end of the pond a stone dam with an exit hole. Plum in the middle of the pond is a smooth volcanic rock, a pumice mound made for displaying wagtails. There is an enormous clump of yellow iris, which must look fabulous in the summer, a ready-made bunch of proffered flowers for celebration or forgiveness.

On a good clear day, the ruin is reflected in the pond, and the afternoon sun will burn out from the water. In the wind of today, the willow leaves shoal to show their silver underbellies, and the pond's surface water is pitted and dipped.

At night, under a moon, the ruin is like the hood thrown back from a cracked skull. The pond is kidney-shaped.

Bordering the pond, the slow-flowing Argenton; beyond the Argenton, a stand of poplars with pom-poms of mistletoe; beyond the poplars, hundreds of tiny fields tied together with oak-scrub hedges, gold-gilted in their autumnal blaze of glory. This is the 'bocage Bressuirais'. Pale Charolais cows drift across the fields.

A wind gets up, but the blank water of the mill pond remains undisturbed. As calm as a mill pond.

I'm in France learning about 'bio' agriculture, and pondering a possible diversion into lavender and wine.

Talking to the mill-owners, a retired couple supplementing their pension with a *gîte* business (we're staying in 'The Mill House'), it turns out that it is sixty years since the mill wheel turned. The then owner fell between the cogs, and was crushed to death. So, rather more Zola than Byron. The *moulin* was used to shred tobacco, as well as – accidentally – human flesh.

Later: five male mallard, two females. The males in their finery, performing the sea-horse rearing rituals of courtship with Busby Berkeley synchronicity; the two females monarchically unamused.

13 NOVEMBER: Two moorhens, feet up in the water, fighting, emitting a stuttery *kik-kik-kik-kik*. These are girls kickboxing over boys. For birds with outsize lily-pad feet, moorhens are surprisingly agile; they can climb trees. After a few minutes of fighting, the moorhens break off, to opposite ends of the pond, where they pick-peck for food with the focused eye of chickens.

To the side of the mill, slow flows the Argenton, with a *moulin* every half mile. Charolais cows come down to drink, peculiarly ethereal for beef breeds.

The next day: I sit still on the balcony, and a kingfisher comes for me, perching on the rail two yards away.

The warmth is faint, but acceptable; a mallard female sails down the Argenton with ten ducklings astern. Whether she is unseasonally late, or unseasonally early, she is out of time.

Later, I walk up the lane above the *moulin*, where a gate to a field with an *étang* – a pond – is open. (It's a real, galvanized gate; as opposed to the barbed-wire-and-stick contraption French farmers usually have.) The *étang* is almost tropical; gorse, a sparking of trees (poplars), a gentle gradual beach, golden sand. The non-paradisiacal notes are a carp swimming in circles, its fin showing, shark-insolent, and the whale-skeleton of a poplar, which fell into the water another age ago.

The Tuileries, Paris
18 NOVEMBER: On the TGV from Angers to Paris Montparnasse, I look out of the train window. On a pond, alone, a swan.

Walking around Paris, we stop at a site of family mythology, the Tuileries in the 1st arrondisement; Penny and I came here with the children on our first ever family trip abroad. At the round concrete ponds – they are raised by walls, two feet above ground level – of the gardens, the toy-yacht stand is closed up for winter. (There is a perfect, immortal painting of the ponds, Picasso's Post-Impressionist *La Bassin des Tuileries*, 1901.) Today, in the stead of wooden playthings, birds are on the water; ducks, gulls, moorhens. In the background, the Eiffel Tower, and a Ferris wheel; the gardens are bordered by the Rue de Rivoli, and its marble-halled five-star hotels.

For a close-up of the secretive moorhen, head to the city, where it becomes unnaturally inured to human presence and forgoes the nervousness that earns it the name 'Skitty' back home. Within touching distance, two moorhens on the ramp into the water fight for position with a pigeon. On the green bench next to me, a small boy in a woollen coat buttoned to the neck encourages the birds with shadow boxing.

The Farm Pond, Herefordshire

21 NOVEMBER: When I went down to collect the dead sheep the wind was already rising. This was about ten in the morning, and the sky was the grey of mould.

If you work outside, you know that wind is not just wind. There's the wind that is like a wall; the east wind with a Stanley-blade edge; the ghoul wind with thrusting arms that reaches in and move things.

The wind on the day of the dead sheep was from the north. A leaf-stripper. As I edged the Jeep Cherokee down the bank of the meadow, I thought about the wind: 'At least it'll dry the ground.' The weather glass, half full.

Our Jeep is a proper off-roader, as in 4x4, as in permanently SORN, as in exhaust attached to chassis with an adapted wire coat hanger, and kept for jobs like this, with the tyres slightly deflated for grip.

As I neared the dead Hebridean, I failed to notice the crow on its head, but, in fairness, they were a colour match. Black on black. Then I saw the crow, and saw it was stabbing. So tempting is eye of sheep the crow did not desist pecking despite the approaching car. Stupidly, I parked the Jeep nose into the wind, which made opening the door impossible, so I had to turn it around.

The extra minute was all the time the crow needed to finish its crude dissection, and as I opened the car door with wind-assist the crow flew, the delicacy of sheep's eye in its beak.

There's an old farmer's joke: Q: How do you tell if a sheep is sick? A: It's dead.

The Hebridean had been strolling around as happy as Larry the lamb the night before; by seven on the morning of the wind it was on its side, as life-full as a slab of mutton in the butcher's window. The head was arched back, as though its final moment had been protest or prayer to heaven.

Sheep usually go to cover to die; they hope for safety from predators, but their sanctuary is their graveyard. My little Hebridean was no exception; she had died beside the hedge, below the willow, which was now weeping leaves, thin and helpless, in the wind.

I examined both ends of the sheep, as you do, and there was no obvious sign of disease or illness. The hollow eye cavity stared dully at the sky.

The coroner's verdict as to cause of death? One of those things.

It was only then I realized how truly raking the wind was. The rest of the flock had gone as one thing, a single entity, below the rise, their backs to the wind.

Leaves from a black oak a field away tempested around me; every tree in every hedge and copse was being defoliated, screaming, down to the bare branch before my eyes. The world was in motion. Nothing seemed fixed. The wire stock fence along the wheat field moaned; a giant lyre played by the gale.

Looking over at the pond, every leaf from the alders on the north bank was being torn off and laid on the surface of the water, as thick as carpet. Those million leaves would sink, and fill the pond from below. Trees: they assist the suicide of ponds.

The weather glass, half empty.

I started pulling up the Jeep's boot door; the wind took it and finished the job. The empty white feed sacks I'd laid down to protect the boot floor from leaking ovine effluent swirled to the front, as though wishing to escape the duty.

There was no point in re-assembling them. Not in that wind.

Hebrideans are not heavy sheep, but they are heavy enough. The knack of lifting a dead sheep is to grab the two legs of the underside, bend one's knees as though weight-lifting, and hoist the carcass on to the thighs, then thrust the sheep forward as one stands up.

It was as I was closing the boot I heard yelping above the storm; for a moment I thought the sound came from the sheep, risen like an ovine Lazarus.

Then, on instinct, I looked up and there were the geese in a determined V, navigating true south like an arrow.

Greylags. They were low in the sky, and they were yelping as they flew.

The 'hounds of heaven', the country writer BB called them.

Nothing electrifies an English farmland scene like geese overhead; they trail wildness on their wings. I watched them go on their way, as big as dogs, to their wintering grounds on the Severn.

Every creature has its element. The geese were fine and dandy that day; they belong to the wild wind. But the rest of us were struggling. The fanning kestrel, who requires mastership of the wind and not obeisance to it, was nowhere to be seen; driving up to the house, with the wind harrowing the long puddles in the winter wheat, the fox was as low to the ground as a snake.

Bits were breaking off the ash trees behind the barn; the wind of winter was sorting out the dead wood of the year.

The tarpaulin over the spare hay stack billowed in imitation of a green parachute.

I had already phoned the knacker's wife, who had diverted me to Mr Williamson's mobile's messaging service.

As I drove up, he was reversing the trailer of his Daihatsu Fourtrak on to the yard; Mr Williamson swears by his Daihatsu, which is colour-coded the same navy hue as his Dickies boiler suit.

His trailer was already more than half full. Dead cows, pigs and sheep. A single goat. All of them stiff, with sticky-up legs, like a dumped collection of toy plastic farm animals.

We threw the Hebridean in the trailer, the way you swing someone into a swimming pool, a person each end. Rob Williamson got into the Daihatsu cab to write the necessary documents. I crouched down by the running board to write the cheque.

We tried to say goodbye but the wind took away

the words, so we were left open-gobbed like fish, beings out of our element.

23 November: By day, the white winter sun is held in the cruck of an alder like a magnifying glass burning a hole in the water; at night, the water is moon-charmed.

It is at night that interesting things happen at the pond; when the four-footed creatures leave their lairs, the macabre owl twists out from its hollow tree in the woods to begin its day.

One's nerves must be strong to remain alone by a pond after darkness has fallen. There's a wind, but the bank of the pond is a parapet against its infiltration; the surface of the water is barely roughed. I loiter like a hunched heron against the alder, its bark coarse, a maddened skin; I shuffle slightly, to find the comfort of the same tree's draped moss coat.

I smell the vixen before I see her; she comes out of the shadow of the hedge, into the moonlight of the field, to meet the darkness again at the far edge of the pond; she aligns precisely with the thin shadow of the winter alder, on which remain only the last orphan leaves.

The vixen laps the water. For a moment I am

tempted to illuminate her with the torch, but it would be a profanity. Then she is away on her night hunt.

I've kept a mental list of the winter animals that come to this waterhole: rat, badger, fox, and the oddities of a single Muntjac deer and a grey squirrel.

The night at the pond is quiet, anticipatory. The tawny owl utters its first screech, and suddenly out from among the stars the fieldfares come *clacking*, tiny black shapes, living silhouettes in the sky.

24 NOVEMBER: Mist makes even a small pond like this mysterious, and the world is slow to take shape. There is a certain comedy in checking white sheep in fog; two ewes are apparently missing, but I eventually find them down in the drizzly sedge of the pond, chewing beatifically.

Poor old sheep, never a good word about them from 're-wilders', but here they are conservation grazing, cutting back the sedge (*Carex riparia*). We go wild for wild animals, yet disdain domestic ones. With their mouth-shears the sheep have already cropped the year's shoots and suckers of alder which, if left unchecked, would smother the pond to death. Trees would take over the world if they could, and have the world in their shadow.

Standing here at the pond, I realize that all the things outside I love are agriculture's compromises with nature: traditional hay meadows, retro arable fields, husbanded woods. And farm ponds.

In the afternoon: a male mallard skid-lands on the water, immediately brightening the pond with his iridescent green head. Just as the musician is interested by Philip Glass's minimalist *Akhnaten* as well as Wagner's big-bang *Der Ring Des Nibelungen*, the naturalist in winter finds beauty in small things. Such as the sheeny green head of the mallard drake.

27 NOVEMBER: The days of clear skies and frost have put a centimetre of ice over the pond. My hands shade my eyes from the sun, the ice-skim looks solid enough to walk on, a white floor. I skip some rattling stones across the ice; then throw one high; it lands, with the echoing ring of doomsday reaching across the hard fields and up the hill.

Under my weight, the ice squeaks, my courage melts. So instead of being the first man ever to traverse this arctic floe I stand on the firm bank and pile-drive holes in the ice using the end of the metal sheep trough. Brown water slops up, ugly and unseemly. The open water is for Skitty and Big Bird to feed, for others to drink. Held to the sky the smashed ice-panes are entirely transparent, their only imperfections a minute spherical bubble or two. Only in Regency windows have I seen purer, clearer glass.

In one piece of ice, there is entombed a button-black whirligig beetle (*Gyrinus natator*), but whether it was frozen in life or death I cannot say.

Withington, Herefordshire
The Remote Past: Ponds: They make for reflection. My very first clear memory is of being bitten in the face by my great-uncle Willi's sheepdog; I was in a pushchair at his sheep farm on the Gower, staying with him and Great-Aunt Kathy while my parents holidayed in Venice.

This was one of my parents' periodic attempts to patch their marriage; it failed, and by the time I was seven they had separated. After the break-up I chose to stay with my father and the dog of my own he had given me, Honey, a golden retriever, plus her accidental son Rover, a black Lab. My family consisted of other animals by then: a tortoise, the Rhode Island Red chickens I took for walks, and a quarter share in a grey pony called Owen. (I had a back leg; such is life.)

Horses were a big thing in my childhood. At St Paul's C of E primary school I went 50–50 with

Malcolm Thomas on purchasing *Horse & Hound* each week; I was seven, and had only just learned to read after a huge, if sternly delivered bribe from my father. ('You can only have the *Thunderbirds* EP *when* you can read': next day I could read.) My father owned a business supplying the catering for national hunt meetings; so a great deal of my childhood was spent watching horses go around Sandown and Kempton Park. Hors d'oeuvres and horses as I came to think of it.

At this time we lived just outside Hereford, the River Wye to the bottom of the long gravel drive, a farm next door, and stretching away down towards Ross the sort of meadow-and-hill pastoral scenes that English men and women paint, write poetry about, or die in wars for. In some parishes the field pattern was the same as it was in the thirteenth century, when my mother's family arrived in Herefordshire.

My childhood Herefordshire was an antique land.

My 'pets' consisted of more than those I officially owned. Every apple tree in our orchard was mine, as were the house martins that nested under the eaves, the cabbage white butterflies on the vegetable patch, the bats in the barn, the treecreeper that climbed the pear tree on the front lawn, professorially peering into every crack in the bark. I was besotted with them all.

There seemed no end to the birds in Herefordshire in the 1970s.

My father, who had served in the Fleet Air Arm during the Second World War, believed in 'benign neglect' when it came to the care of his son. Luckily, the parents of my friend Tim, who were the same Second World War vintage, believed in the identical parenting philosophy, so Tim and I spent weekend afternoons wandering Tupsley Quarry and the Wye Valley with bits of fishing line to catch eels, and out-size Boots 10x50 binoculars to spot birds. The latter were entered on the checklist at the back of the RSPB diary from WH Smiths on Commercial Street that was my staple Christmas present.

What to do about the boy's schooling became another argument between my parents. At one stage, their preferred solution was to send me to a girls' boarding school outside London with which my father had some connection. I managed to avoid that fate by being farmed out to relatives, and employees of my father, so I could attend day school in Hereford until I was a teenager.

The 'farming out' was literal in one sense. When my father's work took him away to Sandown, Kempton Park, Wincanton and those other racecourses that run through my head like a catechism, I lived with my maternal grandparents for weeks, sometimes months, at a time. Joe and Margaret Amos were hop farmers in east Herefordshire. Although the picking of hops had become mechanized, not much else had been touched by modernism down on the farm in Withington.

Neither had the locals. Looking back, the characters of my childhood have a Fellini eccentricity; one of the great losses of the countryside is the country person, now replaced by middle-class executives who drive a Lexus to work in some city. One of my farmer uncles – who grew the blackcurrants for Ribena over by Tenbury Wells – had a fatal love of fast women and slow racehorses. Our gardener, Keith, was the local bobby; he would sign on at the station, cycle the three miles to our house, do a day's gardening, then cycle back to the station in time to sign off his police shift. My stepmother, who was intimidating for her brilliance at all things (from RADA gold medal for dancing to captain of the golf club), was also a bridge master; her expertise at cards meant that she cleaned up at the Christmas whist drives in all the villages of south Herefordshire. I had to go along as her bag carrier, and packer of turkeys, chocolate and sherry into her orange Citroën Deux Chevaux. We lived life on the high at Christmas.

Most eccentric of all, however, was my grand-parents' next-door neighbour in Withington, a bone-thin aristo who would emerge from her lane-side garden to grab me by the arm, and warn, 'John, never, never trust anyone who does not have dog hair on their sofa.'

I was terrified of her, although she really should have been alarmed by me, given I was the only punk in the village. (Back then, I thought the lady in the herbaceous border mad. Curiously, the intervening years have proved her largely right.)

In front of my grandparents' house, East View, was a farm pond. I do not think it a rose wash of memory to say it was the perfect farm pond. I can see it now . . . the grass snake which sidles across its surface, the Hereford bull walking out of the bitumen-painted cor-rugated iron barn to drink its ruddy water, drops dripping rudely from his mouth as he eyes me up. I fished there for sticklebacks on hazy holiday after-noons with David Hughes, son of a farm worker.

The stickleback comes in three- and nine-spined versions, the name being derived from the Old English *sticel*, meaning spike, though lingua vernacula knows the two-inch fish also as tittlebat, the sharpling, the sharpnails, the prickleback, the banestickle.

As a child it struck me as funny that red rags incensed the pond's sticklebacks as well as the bull.

Sticklebacks are usually seen in shoals, but in spring the male leaves the 'school', dressed as a dandy with green back, red belly and blue eyes, and picks a part of the pond, which he defends fiercely, as the dog does the bone. When another male nears his fiefdom he charges the intruder with his spines stuck up, and jaws apart. Often this bold display is sufficient to send the intruder on his way, but if the trespasser stands his ground the 'owner' upends, as if chewing the pond bottom.

As David Hughes and I discovered to our childish delight, a trespassing male stickleback can be imitated by a red rag on a stick, or even a red pencil, waggled in stickleback water.

When not patrolling his patch, the cock stickleback builds a nest by scooping a shallow bowl out of the pond bed, and pressing small plants into it. These he glues together with fluid from his kidneys until he has constructed a tunnel only marginally bigger than his own body. After nest-building his colours become brighter yet, and he swaggers around his territory, hoping to attract a female to his abode. If successful, she lays eggs in the nest. In one of nature's odder connubial arrangements, the hen stickleback departs the home, leaving the male stickleback to care for the eggs by fanning his tail to send a flow of fresh water over them.

He is a good father. The young, when they hatch, are raised solely by him. If any of the sixty to eighty young stray from the nest he swims after them, catches them in his mouth and returns them to the paternal fold.

What was the name of David Hughes's dad, with his bright Army-green Series II Land Rover? I've lost touch with David, and all the adults have gone to death and dementia.

The classic farm pond on the yard, such as that at Withington, with its dabbling ducks and drinking cows, barely exists (if at all) outside the pages of kids' board books, Eric Ravilious' paintings, our old farm in the Black Mountains and the poetry of John Clare. Recently I treated myself to the Delphi *Complete Works of John Clare* (£1.50), which is full of verse snapshot records of the Georgian farm pond and its utility: 'While ducks and geese, with happy joys, / Plunge in the yard-pond brimming o'er'; another poem: 'The horses are took out the cows are fed / and lame old dobbin to the pond is led'; and another: 'And ducks and geese that clamorous joys repeat / The splashing comforts of the pond to meet'; and another: 'Round the pond the martins flirt, / Their snowy breasts bedaub'd with dirt.'

Clare was a lover of ponds, and a worker of ponds too. He worked at Burghley House 'scouring out fish ponds' – that is to say, clearing out weeds and sludge in order to keep the water fresh, and free of the infectious effluvia known as 'den' or 'dain'.

Clare was the godly voice of the small things in the English countryside. Not the grand view, but the intimate and the up-close. So he was always nesting and ponding.

He was still writing fondly of the small still waters in the autumn of 1841, after coming home from the asylum in Essex:

Closes of greensward and meadow eaten down by cattle about harvest time and pieces of naked water such as ponds lakes and pools without fish make me melancholy to look over it and if ever so cheerful I instantly feel low spirited, depressed and wretched – on the contrary pieces of greensward where the hay has been cleared off smooth and green as a bowling green with lakes of water well stocked with fish leaping up in the sunshine and leaving rings widening and quavering on the water with the plunge of a Pike in the weeds driving a host of roach into the clear water slanting now and then towards the top their bellies of silver light in the sunshine – these scenes though I am almost wretched quickly animate my feelings and make me happy as if I was rambling in Paradise and perhaps more so than if I was there where there would still be Eves to trouble us.

An altogether different kettle of fish from the farm pond at Withington were the rushy cow ponds at the bottom of the village, where a pike was rumoured to live. Slippery, fearsome, mysterious creatures, pike.

David and I fished for it fruitlessly, and half-heartedly, with lures borrowed from my father's fishing kit, which he kept in a beige canvas satchel. The cause of our ambivalence was that, pure and simple, we were terrified of the leviathan of the relative deep.

According to piscatorial science, pike are at their most voracious when the temperature of the water is rising through 40–42°F (4–6°C). They are at their most sluggish in autumn, and in cold periods can lie low and lethargic in the pond for days. David and I attributed our failure to catch the Withington pike to the fact it was June, thus practically autumn already.

The Farm Pond, Herefordshire
29 NOVEMBER: I think the thing I like most about November is the feel of the gun barrel against hand,

metal against flesh, which is a sort of metaphor for the first month of winter, when the last heat of Aestas meets the cold of Skade. Before you stop reading, I should say here that I hardly ever shoot anything, and only then for the pot. It is just that carrying a gun makes one a hunter in the landscape, recasts one as a predator in the countryside's web of life. And death.

I am out this morning with the Baikal .410; the .410 is the poacher's gun, particularly in the folding version, hidden beneath a long coat, or in the case of Tom, the village milkman when I was a child in the 1970s, under the vinyl seat of the van for a passing pot at a pheasant. In Herefordshire we regard this as a proper drive-by shooting.

The rain is thick this morning; it is the sort of dollopy rain that collects on top of the cap before trickling down the back of the neck slowly. Like sweat.

Not even the collar of the Barbour Beaufort buttoned tight keeps the rain out. On the contrary, the collar is a funnel that directs the rain down my spine, so I am soaked to the backbone.

The earth aches with sullen cold.

I haven't taken the dog. Even the dense coat of a black Labrador is permeable to such downpouring. Besides, there are times when you want to be alone in the element of rain.

First off, I skirt around a half acre of millet we have down as a conservation crop. My every footprint presses into the earth: a mould that fills instantly with blancmange-pink water. A solitary and bedraggled cock pheasant stumbles out of the bowed stalks. For a moment the pheasant looms in the deluge, copper-brilliant, before fading away.

I don't pull the trigger, though I want his hot flesh, because the onset of winter is when one craves meat. The country festival of St Martin, celebrated on 11 November, was when the cattle were slaughtered for 'Martinmas beef'. We need to put on fat for winter, just as the animals do. The Elizabethan poet Edmund Spenser conceived of November as the month of glutton, 'grosse and fat / As fed with lard, and that right well might seeme'. The old people of the isles knew the need for slaughter in November; the Anglo-Saxon name for the eleventh month was *blotmonaþ*, 'blood month'.

Anyway, I resile from shooting the pheasant due to the pity of it. He is too wet to flee. And I know him and his bold white collar. All the summer long he has aboded my fields and copse, amusing me with his mincing Ming Emperor arrogance.

The rain – and this seems impossible – becomes heavier, pulsating towards me in vertical, drenching curtains. I can scarcely breathe. There is no view. Drops of rain blur the windscreen of my eyes; and lashes are less than wipers.

Head down, I bend into the rain, almost hag-doubled over. Rain makes geriatrics of us all. Want to know your future? Go into a deluge. The shape you assume then is the shape of your end days.

Anyway, on autopilot, I slide-slosh down the bank meadow. The Baikal, held horizontal and low, is comforting, like holding the hand of a sister.

On the flat land before the brook, where all the water of the last week has congregated, drowned worms, in dead white Ss, float in the circular hoof marks of the cows, miniature pools an inch deep. The cows themselves have long been driven to higher ground.

The rain beats on the weakening shield of my jacket.

The flat land is under three inches of water; the rain has created a new pond. Three male mallards have already taken imperial possession of the 'flash'. On my approach, the ducks evaporate. Wild duck in the air are atavistically exciting.

Place has memory, does it not? Overlaid on the scene is the picture of another man with a weapon, awed by the speed of wild duck on the wing. A Neolithic huntsman with a spear.

Again, I fail to shoot.

Across in other fields, the water has gathered in the low places, memorialized as occasional ponds by the sedge, there the year round. And so water re-finds its familiar haunts.

In the neighbour's winter wheat field, there is a definite human-created depression, circular, about ten metres in diameter, now sated and satisfied with surface water. It's an old farm pond, that has been ploughed under. A 'ghost pond'. The story of the rise and fall of the pond has many villains, including intensive agriculture. In the United Kingdom after the Second World War, the intensification of agriculture led to the draining or filling in of small wetlands. Modern machinery meant that a pond that had served as a watering hole for livestock for many centuries could become dry land in a matter of hours.

Anyway, these were some of my thoughts this morning on a walk in the rain.

There are no innocents when it comes to ponds. One contributor to a Mumsnet thread on 'filling in a pond' writes:

My main job that I have left to do before it is born – I have rather a large pond to fill in . . .

45

On a MoneySavingExpert.com forum a grand-parent exclaims:

> I have been draining a pond in my new garden (have young grandchildren and wanted the space) and am horrified at how much filling it will cost . . .

This is the pond cast as enemy, contrary to its function as the child's aquatic wonderland.

10 DECEMBER: I found the lump yesterday morning. She was leaning forward, taking a drink of water, and my hand was on her neck, caressing her.

And there was the lump. It was dark, so I felt further down. Another lump.

It is what every farmer fears, a cow reacting to the skin test for bovine TB.

Oh, the principle of the TB test is simple enough: each cow is injected in two places in the neck by the vet, first with an antigen from bovine TB, secondly with an antigen from avian TB. Seventy-two hours later, said vet comes out to check for bumps. A 'reactor' cow, one carrying the disease, will come up in bumps. Which is a death notice, delivered by a green tag punched in the ear.

Except . . . the assessment of bumps is far from simple. It depends on how precisely big is the bump. Then, is the avian or the bovine bump bigger? Some delicate, wilting-violet cows will bump to the steel needle, let alone its active contents.

The merest hint of a bump on a cow, and I panic. Reason flies out of the window. I fluster. I cannot remember even whether the key skin prick, the bovine one, is top or bottom.

Every cow has worth, but in a small herd of ten like ours every cow has meaning too. Plus, I confess, a name. In the case of our Red Polls, these by family law begin with 'M'. (The book of baby names has been well plundered over the years. We've done the spectrum from the haughty Margo – think a cowy version of Penelope Keith in *The Good Life* – to the buxom Mirabelle, a natural to advertise butter.)

Miriam is a particular favourite, with her fluttery eyelashes, and pawky humour. A couple of years back, she kept escaping, and I was perplexed utterly until I caught her red-hooved. Her modus operandi was to squiggle on her side under the single-strand electric fence.

I wouldn't want to lose her. Really, really wouldn't.

I tell you now, the three-day TB test is hell. I don't sleep before it, don't sleep during it.

Today is D-Day: the day the vet comes back to check for 'bumps'. I got the cattle penned for examination just after first light, and getting the girls into the pen is sufficiently stressful for me to pop 75g aspirin pills. (I read somewhere once that aspirin mitigates strokes. Or heart attacks. One or the other.) It is stupid, but one does not want to be the cowman whose failure to pen cows is tittle-tattled by the vet down the line to the neighbours.

I took no chances. I led the girls into the pen with a shaken bucket of cattle cake, me running faster than anyone at Pamplona, because being caught up by five tons of milling beeves is interesting to say the least. The girls in the pen, I two-armed vaulted the side, nipped back around and closed the gate on them.

Works every time, except the time I failed to clear the pen's side, fabricated from motorway crash barriers, and broke three ribs.

A really great cowman, like our old neighbour Jack Williams over at Abbeydore, can pen cattle, even burly Limousins, by hypnotizing magic. He sings them in.

It is almost ten o'clock now. The vet is due, and I'm by the pen, paralysed with worry, mouth sticky and parched, as when you receive a Communion wafer.

The mist is low, axing off the top of trees beside the pen.

There is no compassion in a tree; on a dolorous day like this their unthinking insensitivity, their self-regarding composure merely irritates.

Somewhere off in the hedge, an unseen robin drips sad soliloquies.

Lower still comes the mist to lie on the backs of the cows in white shrouds.

My wife, Penny, comes out with a message: the vet is delayed. Another half-hour.

I have to get away, and take myself by the scruff of the neck to the pond. It's a mistake; ponds reflect moods, and as I crouch foetal on the bank I see the milky water is full of immense melancholy. Around the pond, the grass and the sedge is razored to the skull, and the wind has sandpapered the rim of the east bank to a sharp edge. You can see the shape of the land in winter; it is a body, and the ponds are its wet eyes.

Nothing stirs, nothing moves. Even the moorhen is silent, and keeps to her secret places under the willow. Skitty is the soul of the pond in November. I miss her.

Like many a farm pond, ours is entirely artificial, made three hundred years ago by cleverly damming a spring to create a watering hole for cattle, and a place to grow carp for fish on Friday. There are still carp, though I try to fish them out periodically. They are the destroying devourers of diversity in the pond.

Unseen, down in the oil-thick water the fish lie torpid.

I suppose one reason that ponds have never had much vaunting by wandering Romantic poets is their quietude; there is so little sound to riff with, unlike the fluting brook, the percussive waterfall, the gale-shrieking sea.

The winter notes of the pond are a deeper music still; the pond stores the sounds of slow, old England; the clop of horses' hooves, the clang of swords, drunken men singing folky songs on the way home from the pub.

I hear their youthful voices, I see the sweat on Dobbin's flank, I feel the cold metal in my hand . . .

The thirty minutes is up, and I trudge slope-shouldered back to the cow pen, and the standing, staring cows, their breath making more mist.

The tweed jacket era of James Herriot is long since gone. Bob Minors looms towards me in green waterproofs, a doppelgänger for an alien in *Doctor Who*, c. 1980. He is accompanied by Bluebell, our black Labrador, who likes to keep an eye on interlopers, and escort them around the premises.

'Any problems?'

I grimace. Then push the internal gate behind the cows, so they are pushed into the narrow metal-railed corridor of the 'race'. (An odd noun for the place for the queuing of animals.) To touch, the gate has condensed the chill of the centuries in its bars.

Cows are hierarchical. Miriam is mid-herd, number four.

Bob leans over, runs his hands over the injection sites on the cows' necks. He reaches Miriam. 'Ah,' he says. And gets, from out of his tool box, the callipers.

He squeezes her warm vital skin and pinches it with the steel callipers. Once, twice, thrice, he measures the bumps on Miriam's neck.

Somewhere across the valley comes the coffin-closing croak of a raven.

'Close, but good enough,' says Bob.

I do not understand, and ask him to repeat.

'She passes.'

The world blurs. He runs his hands over the remainder.

'They're yours for another year. I'll see myself off. You let them out.'

I famously only swear when I'm happy. '****! ****! ****!'

I release the cows from their confinement, and me from my anxiety.

We run around, the cows and I in the mist, laughing.

Afterwards, I swing into the tractor seat, as familiar to me as the desk chair is to the office worker, and head for home. Bluebell jumps into the transport box on the back of the Ferguson. We totally cane it, the throttle lever pulled right around, the exhaust on the bonnet blasting like Krakatoa, up into fourth.

I cockily lean into corners like a motorcyclist. The wind whistles past. Bluebell barks with excitement.

Flight Pond, Michaelchurch Escley, Herefordshire
Sometime in the near past, c. December 2006: I made the flight pond by accident. Excavating some foundations for a field shelter, I found the water table too high, made the best of a bad thing (a weekend's hire of a mini-digger) and dug a pond to shoot wildfowl over. The heavy, red clay was as impermeable as plastic, and to keep the water topped up I re-channelled a drainage ditch. The slightly moving water reduced the problem

of weed growth. I did everything right; left a mound island for nesting in the middle, sloped one bank for dabbling ducks (which require water no deeper than 40cm to search for food, and easy access to the bank to preen), excavated two metres deep one end for diving ducks. I planted reeds in the ditch to filter silt. Come August I started a quotidian scattering of barley grains. There was no need for a hide; I could cooch in the ditch, up to my chest.

The ducks came, mallard and teal. I could see them both on my white china plate, drizzled with cranberry sauce, and a zigzag of pea puree – the nouvelle cuisine style that is obligatory these days. I expected the fowl to visit by night, which they resolutely refused to do, always coming in at dawn. So I settled for the shooting of the morning flight. I fancied myself a wildfowler.

One October morning, I got down in the ditch with the 12-bore. The darkness was so dense there was no weld between sea and sky. I had the black dog with me, she perched on some sacking on a ledge, mountaineer-style, and we waited for the dawn.

The frost on the grass, the immensity of the star-studded night, these filled the senses.

There was silence, except for the distant *hush-hush* of the Escley brook. There was the fowler's pleasure in contemplation and lonely places. And the Black Mountains of the Herefordshire/Wales border are among the last wild places in England. The mountains were a wall of dark, except for that farmer in Cloddock with a fluorescent orange light over his yard; a street-lamp intrusion.

Wildfowler's Confession: if fowling was cushy, fowlers would not do it. Part of the allure of wildfowling is that it is Man v. The Elements. The metal of the trigger was so cold it burned my fingers.

Somewhere out there a barn owl screeched, with all the desolation of the mountain night in its call.

A blade of light appeared in the east, and the brightness grew over the land, which resolved into muted winter hills and fields.

The two teal came in, comets falling, and as I swung the shotgun up, a vision came into my head: an oil-painting kit I was given for Christmas, aged twelve. The painting was of a teal drake; how I laboured over my artistry. (I still have the painting in an attic somewhere.)

I pulled – did not take – the shot, and I have been pulling shots on teal ever since. I was overcome by the delicacy of a memory, and the delicate beauty of teal, our smallest duck.

The pair landed, then saw me. The collective noun for teal is a 'spring', which is apposite; they sprang out of the water, jacks in a box, back into the sky. Like dawn light on water, the emerald stripe of the teal's head is a great thing.

I never shot a duck on that so-called 'flight pond', and it turned into a duck pond as you'd understand it. Snipe loved it too, especially as the water diminished towards summer, leaving the fleshy ooze naked, which they printed every dusk and dawn with their feet.

The Mill Pond, Argenton, Western France
12 JANUARY: Snow, about two or three inches deep. Everything white and black, the birds picked out in black against the background. I feel Victorian walking up the lane, a wight alone, an illustration in a pen-and-ink drawing; the quiet, the lack of traffic. Five mallard fly off from the *étang*, with its thin cold water.

Of course I have a dog with me: I am not a rambler. I am a walker: I like to belong, or pretend I belong.

Crossing the stone bridge over the river, I notice two cormorants on the bank, oil-slick shiny, fish-oil glossy.

Later they fly over the mill pond, which is iced to

the hilt; they are the pterodactyl bird; under their travelling shadow it is Late Jurassic.

> *The glutton Cormorante, of sullen moode,*
> *Regarding no distinction in his foode.*
> Bibliotheca Biblica, 1725

Cormorant, from the Latin *corvus marinus*, meaning 'sea raven', and always sinister in literature, as well as providing a synonym for gluttony, like the gannet. In Book III of *Paradise Lost* Milton chose the cormorant as simile for Satan as he plans Adam and Eve's downfall:

> *. . . on the Tree of Life*
> *The middle tree and highest there that grew,*
> *Sat like a cormorant; yet not true life*
> *Thereby regained, but sat devising death*

English royalty was less disposed to flights of fancy; James I kept cormorants on the Thames at Westminster. These semi-domesticated birds would fish for the Master of Cormorants, a member of the royal household.

The Farm Pond, Herefordshire

20 JANUARY: Any leftovers and clutter from summer now being stripped out by the wind; birds tumble akin to clothes in a giant washing machine. The dog fox, slinking low to the ground out of wind's way, circles the pond, looking for ducks in broad daylight.

The wind brillos the surface of the water. Low, repeated call of robin all around, strung from tree to tree. The pond has many moods; today it is steel, and unforgiving.

21 JANUARY: Snow. Wet, hard snow, which came down on a northern sky just after dawn.

The sheep have retreated to the lee of the hedge, to stand like disconsolate spectators at football. Only a single cock pheasant, circling the hay rack, saves the field from loneliness and white monotony.

Snow. Sticky snow, that forms plates under the feet as they lift up.

Snow. Hissing snow that blanks out sound, so there is only the steady shimmering of icy pellets hitting me, the snow crust, the ice sheet on the pond.

Around the edge of the pond, the repeated triangular footprint of Big Bird.

The ice has locked the heron out of its larder. Following 1963's big freeze, heron numbers dropped by perhaps as much as half as rivers and ponds froze over. So I once again become the happy vandal, and smash ice. Held to the light it is a slice of ruby; the sheep had been in the water previously, at the sedge, churning up the red clay of Herefordshire, made 400 million years ago when the world was swamp.

At night, when the snow stops, I hear the forlorn incantations of an owl, far away, such bass sonorous sounds as the frozen earth would make if I struck it with a metal bar.

The Mill-Pond

The sun blazed while the thunder yet
Added a boom:
A wagtail flickered bright over
The mill-pond's gloom:

Less than the cooing in the alder
Isles of the pool
Sounded the thunder through that plunge
Of waters cool.

Scared starlings on the aspen tip
Past the black mill
Outchattered the stream and the next roar
Far on the hill.

As my feet dangling teased the foam
That slid below
A girl came out. 'Take care!' she said –
Ages ago.

She startled me, standing quite close
Dressed all in white;
Ages ago I was angry till
She passed from sight.

Then the storm burst, and as I crouched
To shelter, how
Beautiful and kind, too, she seemed
As she does now!

Edward Thomas

The Mill Pond, Argenton, Western France
25 JANUARY: Daffodils out; the long tresses of
weeping-willow gold in the morning light, and

embedded in them, long-tailed tits, uttering their silvery sibilant contact calls; on the pond a moorhen pair and a mallard pair, the legs of the males respectively gaudy green and gaudy orange, breed-time bright, luminous like medieval paint; alder catkins, port-wine vivid; breed-time bright too.

The moorhen's feet, big like pond lilies, require it to walk carefully, like wearing too big shoes.

On a walk up the lane in the evening, with the sweet smell of Charolais cows all around, a couple of black darts shoot over, going north, polar direct. Cormorants or shags? Penny: 'I'd prefer them to be cormorants.' The *étang*: a Romantic mirror, all the Madame de Pompadour pinks of the sky softly reflected on its surface.

At night, an eye-burning supermoon. In the morning, the moon's impression remains in the sky; a thumbprint on blue glass.

28 JANUARY: At night, the strange bleating of coypu, like the cry of lost sheep.

In the morning, one coypu, huddled over, miserable in the poplar wood. It's been shot by a local cattle farmer, who blames coypu for undermining the banks of the river.

For a rodent, coypu look distinctly friendly – my daughter has declared them 'so cute' – though the orange teeth are goofy. Their long outer fur is ratty grey but the inner fur is mole-like, dark chocolate and very soft. A coypu wears a dry-suit; the inner layer of hair is so fine no water can reach the skin. The ancestors of the coypu now invading the slow-running rivers and ponds of Europe were brought to the continent between the wars to produce this inner fur, known as nutria.

When swimming, the coypu looks like a drifting log; it shows less back than the otter, and its tail is thinner. Only the hind feet are webbed, and they are considerably bigger than the front ones. They can reach twenty-six pounds in weight.

Coypu show the wisdom of nature; the mammary glands are situated along the side, so that the young can feed while their mother is in the water. It is a quaint sight to see three or four floating at right angles to her, as she lies with only the top of her back and head above the surface. At birth the young are about seven inches long, including the tail, fully covered in fur and able to swim. Breeding may take place at any time of year, and the female rears up to three families a year.

Being strictly vegetarian, coypu do not scavenge as much as rats; their most common foods include the roots of water dock, bulrush, sedge. They also have a taste for farm crops. From the farmer's point of view, they are the criminal perpetrators of crop damage and bank erosion. Pests; forever and always pests.

The coypu has a remarkable trick to escape pest-hunters: it can hold its breath underwater. One coypu decides to illustrate the gift this morning.

Bluebell the Labrador leaps in the thin cold water of the Argenton after a coypu, which dives. Bluebell circles, a canine destroyer seeking a U-boat. Bemused, the dog concedes after a few minutes, and clambers on to the bank. The coypu remains beyond the reach and patience of dog. Like whales and seals, coypu can restrict the flow of blood to the brain, so that the metabolism slows down. Bluebell's coypu could have remained underwater for twenty minutes.

The coypu population in Britain, based on fur farm escapees, reached its peak in the 1950s, when there were believed to be about 100,000 individuals, mostly located in the east of England. Starting in 1981, at the urging of the farming lobby, an extinction campaign was waged by twenty-four government-employed trappers, who baited weld mesh cages with carrots; the captured animals were dispatched using a .22 pistol. By 1989 the coypu in Britain was extinct, the last two, both elderly males, being run over by cars.

In France, though, they remain in their lamented hordes.

Neutral Tones

We stood by a pond that winter day,
And the sun was white, as though chidden of God,
And a few leaves lay on the starving sod;
– They had fallen from an ash, and were gray.

Your eyes on me were as eyes that rove
Over tedious riddles of years ago;
And some words played between us to and fro
On which lost the more by our love.

The smile on your mouth was the deadest thing
Alive enough to have strength to die;
And a grin of bitterness swept thereby
Like an ominous bird a-wing . . .

Since then, keen lessons that love deceives,
And wrings with wrong, have shaped to me
Your face, and the God curst sun, and a tree,
And a pond edged with grayish leaves.

Thomas Hardy

Still Water

The Farm Pond, Herefordshire
31 JANUARY: I go up to the pond, look in, and see the depths of winter.

This is about three in the afternoon, under a grey sky, and the pond is a ring of dull water. The surface shroud of pale scum, with its litter of twigs, is anti-human.

Reeds take the guise of broken swords. Beyond the pond, the smothering mist.

I peer at the unmoving water; there is only the cold decay of leaves on the pond bottom.

Is there anything more desolate than still water in winter?

All the long day I've been in the barn servicing the tractor, and setting the plough for the new season, a hateful job of delicate geometry and big spanners. A breath of fresh air, away from oil and dust, seemed a good idea. But now, in the drab of a mid-January afternoon, I am less sure. Electric lights and a thermos suddenly seem warmly welcoming.

It's the dog who finds the dead squirrel. She paws at the animal's head, so it swivels on the spindle of its spine.

I call Bluebell off; she refuses to come, so I drag her by the collar, and make her sit. She, usually the most placid of dogs, whines her displeasure.

I understand then that decades of selective breeding have failed to extinguish entirely the wolf inside a black Labrador.

In front of the squirrel, which is warm to the touch, is the smashed white-china of a wood pigeon's egg. Above both squirrel and egg looms an alder.

Turning the squirrel corpse over, I find small raised tufts of fur on its grey back. And bright blood spots.

In the mud around the corpse are the paw prints of Reynard, and one faint scuff from a bird's wing.

Sherlock Holmes of 221B Baker Street is not required to establish what has taken place. The squirrel has raided the wood pigeon's nest, but in its getaway has been attacked by a bird of prey, almost certainly a tawny owl forced to fly in daylight because of hard times.

Few British birds of prey will take on a buck squirrel. The tawny has the weight and the necessary equipment, talons in zygodactylous ('paired toes') arrangement, with two claws pointing forwards and two back.

The talons make a spring trap of claws. As the owl's talons slice in, then close up, the prey dies from shock or the puncturing of a vital organ.

But no sooner has the tawny brought the squirrel

down, than an opportunist fox (is there any other sort?) has interrupted the owl. The dog and I have disturbed the fox.

Elementary, my dear Watson.

Nature is a game of consequences. The pigeon eats the acorn; the squirrel takes the pigeon's egg; the tawny seizes the squirrel; the scavenging fox comes for the dead meat.

Or, nature is a food chain, in which the links are death.

Something comes over me, and I can't stand demise and lassitude any longer so I grab a stick and chuck it at the dead pool. The dog follows for the retrieve, and the water comes alive in small waves of excitement.

Ponds have no motion of their own. They need external forces: wind; jumping fish; paddling mallards . . . swimming Labradors.

January, of course, is named for the Roman god Janus, who faced two ways. January is the transition between the old year and the new year. Between winter and spring.

I had hoped to find some sign of spring in the pond. I had sought frogspawn, or the first green blade of rising reed.

There are no such indications, but as the dog and

I turn to go home, a mistle thrush starts up in the top of the pondside willow.

He is also a herald of spring. Since November this mistle thrush has sat in the hollies, jealously guarding his food store. Now, he has emerged to stake his claim to breeding land and wife in song. The yearly cycle of mating and birdsong has begun.

And doesn't the willow on which he perches have a blush of green? The colour of spring appears early in the willow's skin.

In this moment of optimism, I look up at the wood pigeon's chaotic nest of sticks in the alder. Perhaps, rather than being an absurdly late nester, she is an astutely early one, and can divine the coming of spring better than I can?

It is human nature to live on a diet of hope. I go back to work in a better mood. In fact, you might say, I have spring in my step.

So ends the tableau of 'The Pond in Winter (with Squirrel Corpse)'.

Pond terms:
Aquatic – plants and animals that must live in fresh water for at least part of their life cycle.
Bog – a wet, acidic area that normally has no standing

water, usually on nutrient-poor land and dominated by plants liking acidic habitats. Bogs often occur on lowland heaths, but prefer uplands.

Eutrophic – rich in nutrients.

Fen – like a bog, an area usually without standing water, but richer in nutrients, flora and fauna. The most famous fens are those of Cambridgeshire.

Kettleholes – formed in glacial times by the melting of blocks of ice that had become embedded in moraines.

Marsh – wet area that is more or less neutral or alkaline, with plants that are mostly herbaceous, with reeds and rushes dominant.

Pingos – depressions sculpted by the retreating glaciers. There are dozens in the Brecks in Norfolk, the largest density in the UK. 'Pingo' is an Inuit word for 'mound'.

Swallow-holes and sinkholes – a feature of chalk and limestone country, where ponds absorb streams but nothing visible emerges. The mother of all sinkholes was Lake Copais of Ancient Greece, a flat-bottomed depression which contained a fluctuating mere nine miles across.

Swamp – wetland in which the surface is covered with water for most of the year, but not open or deep enough to be a lake.

The Farm Pond, Herefordshire
2 FEBRUARY: Flocks of gnats, clouds of white sheep.

3 FEBRUARY: Grey sky, grey pond. Mirror match. The pond seems to have no depth, as though the surface was solid.

A pied wagtail flutters up from the bank; a pierrot. As E. M. Nicholson observes in his *Birds and Men*, the pied wagtail, a former cliff nester, has become dependent on man because of its need for the insect life of shallow waters and the proximity of animals – farm ponds, in other words. Nevertheless, the pied wagtail has not ceased to regard humans 'with nervousness and mistrust'. Wise bird.

Skitty the moorhen, that other nervy bird, is on a temporary pond at the edge of the wood, looking guilty in her trespassing.

6 FEBRUARY: Snowflakes on long drops through the willow. The ducks on the bank, higher up than normal, heads tucked around into their back; headless feather boulders.

I see that snow does not fall on water but *in* water,

which has an open mouth. Water swallowing water (in another state), eating itself.

After snow, the weather turns, the world melts, ditches stream, my pond runneth over.

At night, more rain, so the morning is a foreign terrain of small ponds in the corners of fields. Waterworld.

Only a week in, and February has honoured its Anglo-Saxon appellation of *Fill-dyke*.

The Mill Pond, Argenton, Western France
8 FEBRUARY: By the cemetery, a single brave dandelion in the verge. At midday I turn and face the sun like a sunflower. The neighbour's cows are at their hay rack; this day synthesizes winter and spring.

On the lane back to the Mill House, I stop beside a newly constructed pond, with stone walls, and a decent quarter of an acre of surface water. France is pondland; the sheer scale of the place, and the long, dry summers, have conspired to make farm ponds necessities. The public tap does not reach, and/or is turned off in droughts. So French farmers harvest winter's water in *étangs*.

The pond on the lane is multi-purpose, and I take student notes. For agricultural use, it has a bloody great pipe coming out of it, to be attached to a bowser; for recreational use, it has smooth, soft sloping lawn-like banks; for fishing, it has carp and a portaloo.

Before I can stop her, Bluebell takes a flying leap in, swims round and round, her tail a rudder. The cold would have killed me, but add swimming pool to the list of the pond's employments.

The rooks have returned to the rookery in the poplar wood by the mill pond. They croak with self-satisfaction.

The Duck Pond, Hereford
13 FEBRUARY: 8.30am, and because the traffic has been unexpectedly gentle with me, I arrive early for my dental appointment. With nearly an hour to kill, I start walking, but not aimlessly; I head towards the nice part of town, the cathedral and the duck pond.

Actually, the 'duck pond' is a hundred-yard rectangular section of the castle moat, one of more than five thousand moats in England bordering castles and forts, as well as private houses and churches – even tennis courts, as in East Hatley, Cambridgeshire. As well as defensive barrier, the moat has been the Englishman's status symbol, and utility, the place to dump sewage and crop fish.

There are white scabs of old snow on the towering castle bank along one long side of the moat; but the view is made pretty by five pairs of mallard in the water, the long gardens bordering the moat, and the reflection of The Fosse, an Italianate confection of a house in the far end of the water built in 1825. The houses at the street end of the moat are Victorian red-brick houses with names like 'Castleville'.

A woman walks along with a Scottie dog, both dressed in smart tartan coats.

This is the pond as urban tranquillity, mental oasis.

I sit on one of the two benches, both accessorized by flowerbeds and tended grass lawns; on the other bench are a couple of rough sleepers, already at the fortified red wine, but pacified by the sun and the situation, eyes closed, nearly asleep.

The duck pond is almost next door to Hereford Cathedral School, a co-ed private school, so the cars parking for prep 'drop-off' are the usual suspects: Audi A4 estates, BMX X5s, Porsche Cayennes, Range Rover Evoques.

Nobody feeds the ducks. Nobody gets the tot out of the safety seat to, chuckling, chuck food to the ducks, feel the bone-electric jolt when the bird's bill strikes the palm to take the food direct, as delicious as putting your tongue across the terminals of a battery.

I'm not going to lie. This is my end of town. I was confirmed in the cathedral; my parents brought me here to feed the ducks; we brought our children here to feed the ducks, buying discounted granary rolls from Bloomers on St Owen Street.

But: we all live in ponds, of money, of geography. And ponds of time. Because, in an hour of watching, nobody feeds the ducks.

Slimbridge, Gloucestershire
15 FEBRUARY (and the remote past):

'I think there is a wild corner in the human spirit that answers the call of the wild geese.'

Peter Scott

As a child I lived in birds, especially waterfowl. So, many of my favoured books featured the big birds of the water: I call to mind Arthur Ransome's *Great Northern*, BB's *Manka*, *The Sky Gypsy*, and Peter Scott's *Eye of the Wind*. His *Coloured Key to the Wildfowl of the World* was my bible; I still have my copy, bought from the gift shop at Slimbridge, the wildfowl centre founded by Scott. The card cover, splattered and tattered, shows my age.

I was a 'gosling', a youth member of Scott's Wildfowl Trust, and my gold gosling badge was among my proudest possessions when I was nine. I even did the wildfowl identification tests at Slimbridge. By eleven, there was, I think, no species of duck, goose or swan I could not identify.

Like Scott, I was in love with wildfowl, which, of all birds, seem to come trailing mysteries, arriving as they do from the other worlds of the air and water.

I don't want to give the impression of a boy without human friends. You don't have to be a kid misfit to love birds. I learned that from Peter Scott.

As a boy I hero-worshipped Peter Scott, son of a hero.

In his own words, Scott was a 'naturalist by the design' of his parents, his father specifically. A few days before he died in his tent in the Antarctic on 29 March 1912, Robert Falcon Scott wrote to Scott's mother: 'Make the boy interested in Natural History: it is better than games.'

She succeeded, beyond the wildest expectation, though Scott turned out to be rather good at sport too. He was a national-level ice skater, and in 1936 won an

Olympic Bronze in O-Jolle dinghy class sailing. Late in his life, 1963, he was the British Open Gliding Champion. Indeed, there was little that Peter Scott failed to excel at; he was the closest that twentieth-century Britain came to a Renaissance Man. As his biographer Elspeth Huxley noted: 'For sheer versatility and ability to succeed at so many activities, his life is hard to beat. Indeed, it is sometimes difficult to decide what he is best remembered for. One can choose between his evocative paintings of ducks, geese and swans flying across magnificent skies, the long series of radio and television broadcasts, especially *Look*, which did so much to introduce natural history to the British public, his creation of the Wildfowl and Wetlands Trust, his role in founding the World Wide Fund for Nature, and the many books, or perhaps it is all of them and more.'

His talent at art came from his mother, either via genetics or environmental conditioning. In her day, she was a well-known sculptress, and almost everybody who was anybody sat for her. 'I was drying my son's hair in front of the fire when the Prime Minister arrived,' she wrote about Asquith's visit in January 1915.

Scott was educated at Oundle, then went up to Trinity College, Cambridge, to read Natural Sciences. But he was already bitten by the bird and painting bugs, and graduated in History of Art in 1931. He quickly made a reputation for, and a livelihood from, wildlife painting; being his parents' son helped open doors. The son of 'Scott of the Antarctic' was a public curiosity, and his mother's *Who's Who* contacts ensured the great and the good visited Scott's London exhibitions. His illustrated book of fenland wildfowl adventures, *Morning Flight*, was an immediate success. By 1939 Scott was making £4,000 (about £158,000 in today's money) per year from painting. The Medici Society sold 350,000 prints of *Taking to Wing*, a scene of shoveler ducks rising from marshland.

At his lighthouse home in East Anglia he installed six pink-footed geese in 1933 – the beginning of the wildfowl collection that was to become the largest in the world. By 1936 his wildfowl collection had grown to thirty-two ducks and twenty geese. It was around this time that Scott had a change of heart about shooting wildfowl. He did not, he was at pains to make clear, develop a sentimental 'dear-little-dicky-bird' attitude but rather 'a respect for a wild creature which was supremely able to look after itself, and which did so by and large, very effectively in spite of all my wiles'. Whatever; he eventually gave up wildfowling. Besides, by 1939 there was a bigger shooting party on.

On the outbreak of Hitler's War in September 1939, Scott joined the Royal Naval Volunteer Supplementary Reserve, 'a special band composed largely of yachtsmen who believed themselves to have an assured place in the war from the beginning of hostilities'. Scott had a good war. He served on the Atlantic convoys aboard HMS *Broke*, the first RNVSR appointment to become a First Lieutenant on a destroyer, before taking command of a steam gunboat (SGBG) in 1941. He took part in the Dieppe raid, and numerous actions against German naval forces in the Channel, later recounted in *The Battle of the Narrow Seas*. Rightly, he got the medals: two Mentions in Dispatches, a Distinguished Service Cross and Bar.

As in peacetime, Scott in wartime was his parents' son. If he displayed the gallantry of his father, he displayed also his mother's artistic sensibility. Scott is credited with designing the Western Approaches 'shadow camouflage' scheme, which disguised the look of ship superstructure. For this work he was appointed a Member of the Order of the British Empire.

Only in his personal life did the polymathic Peter Markham Scott find failure. He married nineteen-year-old Elizabeth Jane Howard at St Mary's Church, Lancaster Gate in April 1942; but his time with Coastal Forces was 'distant and dangerous', and he blamed himself for the swift breakdown of his marriage. Elizabeth Jane Howard bore some responsibility. She was serially unfaithful; among her early extramarital lovers was Scott's half-brother, Wayland. Howard left Peter Scott to make it as a writer, taking nothing, not even their three-year-old daughter, except a half-finished novel (eventually *The Beautiful Visit*, a sort of classic, the first of several). She married Australian broadcaster James Douglas-Henry, then author Sir Kingsley Amis.

Effectively wife-less and jobless at the end of the war, Scott stood unsuccessfully for the Conservative Party in the 1945 General Election. He lost Wembley North by 435 votes. Politics' loss was wildfowl's gain. He returned to birds 'with a passionate delight', and the help of a chance letter from an acquaintance, Howard Davis, a farmer living in the Severn Valley above Bristol. Davis invited Scott to inspect the immense flocks of white-fronted geese over-wintering on the saltings on the south bank of the river.

So in December 1945 Scott met Davis by the bridge over the Sharpness and Gloucester Canal in the village of Slimbridge. They walked down the lane to a wartime pillbox overlooking the estuary. As Scott later remarked in his autobiography *Eye of the Wind*, 'In most places the secondhand value of a wartime pillbox is strictly limited.' At Slimbridge, the pillbox allowed Davis and Scott to view the geese undisturbed.

Only an expert can tell a lesser white-fronted from an ordinary white-fronted (the lesser has a yellow eye-ring). Davis and Scott were experts, and in the crowds of geese spotted one lesser, then another. For Scott, that settled it. It was 'a moment of unforgettable exultation – a turning point'. Then and there he made the decision to establish on the Severn Estuary a centre for the scientific study, public display and con-servation of the wildfowl of the world. On Sunday, 10 November 1947 at a meeting at the Patch Bridge guest house, Slimbridge, The Severn Wildfowl Trust was duly formed, later the Wildfowl and Wetlands Trust.

Scott built a house there, having decided 'that my home must always be within sight and sound of the winter wild geese'.

It was because of Scott, and my father, another wartime sailor, that my first career choice at school was gunnery officer in the Royal Navy.

Scott, a wartime hero, made bird-watching cool. My early association as a young boy with Hereford Ornithological Club suggested that birding was for the meek. Everybody else, save for my friend Tim, was middle-aged, wore cagoules and walking boots with red laces, held admin jobs with the council and had their sandwiches wrapped in Bacofoil inside a Tupperware box (surely one or the other?). To take a walk with HOC was to walk into Mike Leigh's play, *Nuts in May*.

Back then: I wanted dash and danger. I wanted the Royal Navy. All was going well until, aged fourteen, my maths master, Mister Hull, took me aside. He began by calling me by my Christian name, a certain indicator that sad, bad news was to come. Boys at school were always summoned by their surname. Only one other occasion was the rule broken; when I was approached in the middle of French and asked to go to the Head's office; there I found my father waiting to tell me that my stepbrother had been killed in a car accident.

Anyway, Mister Hull explained to me, not unkindly, that my maths was so execrable that the Navy would never let me near a weapons system, with all the calculations of trigonometry and angles. Never. In fact, they would be unlikely to let me near a boat.

This inability with maths thus also destroyed my reserve career choice: veterinary science. (Here at least, I had a sort of recompense by writing, many years later, a biography of James Herriot.)

By my count – admittedly shaky – I have been to Slimbridge twelve times in my lifetime, nine of

these with my father and stepmother. ('Very support-
ive, your parents,' says my wife periodically, a phrase
simultaneously light and loaded.) My wife and I had
one of our first dates at Slimbridge. Such is the power
of love.

I am back today, on a wild goose chase. I suffer
from goose-fever, the desire to see and hear the wintering wild geese. There are other sufferers. Some take
shotguns for it, some take binoculars.

At 9.30am I am first through the doors of all-
new shiny Slimbridge Wildfowl and Wetland visitor
centre, Gloucestershire. There are pens and ponds of
captive wildfowl, many in species-saving breeding pro-
grammes. Today I ignore them in favour of the wooden
hides overlooking the broad, aluminium Severn. We
who enter the gloom of the hides do so with the rever-
ent hush of church-goers.

The Severn Estuary is one of the main wintering
places for wildfowl in Britain. Out on the salt-grass
and the saline meres there are – and it is a miracle to
behold in modern Britain – pintails, shelduck, teal,
mallard, barnacle geese, gadwall.

Lapwings trip along a mud shore. A huge shoal of
wigeon dib at the turf. Shovelers bob on the bodies of
trapped water; saline ponds. There are greylags too,
only thirty feet from the Holden hide. Seven of them

asleep; two on long-necked sentinel duty. Close-up they are amusingly akin to farmyard geese in their dumpiness.

The geese become restless, and start gabbling. The salt wind increases its severity. A curlew suddenly cries out at the desolation of the afternoon. At some hidden signal, the greylag begin running along the close-cropped grass, then thresh their wings to rise up into the sky.

Ank. Ank. Ank.

Aside from wintering wildfowl, Slimbridge held another attraction for Scott. It was the site of the New Berkeley Duck Decoy.

A 'decoy' is a device for catching wild ducks en masse; a pond is provided with channels called 'pipes', which are roofed over with netting on hoops and screened along the sides; the screens contain spy-holes for observation. The wild duck alight on the pond and are enticed into the mouth of one of the pipes, by trained 'decoy' ducks.

Daniel Defoe gave the classic description of the decoy in a chapter entitled 'The Traitorous Ways of Decoy Ducks' in *A Tour Thro' the Whole Island of Great Britain, c.* 1724:

The art of taking the fowls, and especially of breeding up a set of creatures called decoy ducks, to entice
and then betray their fellow-ducks into several
decoys, is very admirable indeed . . .

Decoy ducks are naturalized to the pond, by being
hatched and bred there. Fed on corn they remain in
situ; when wild duck fly in, the decoy man throws corn
into the water . . . The decoy ducks greedily fall upon
it, and calling their foreign guests seem to tell them
that they now may find their words good, and how well
the ducks live in England . . . When the whole quantity are thus feeding greedily following the ducks or
decoys and feeding plentifully as they go, and the decoy
man sees they are all within the arch of the net and so
far within as not to be able to escape, on a sudden a dog
which 'till then he keeps close by him, and who is per-
fectly taught his business, rushes from behind the reeds
and jumps into the water, swimming directly after the
ducks and (terribly to them) barking as he swims.

Immediately the ducks (frighted to the last degree)
rise upon the wing to make their escape, but to their
great surprise are beaten down again by the arched
net, which is over their heads. Being then forced into
water, they necessarily swim forward, for fear of that
terrible creature the dog; and thus they crowd on,

'till by degrees the net growing lower and narrower, as is said, they are hurried to the very farther end, where another decoy man stands ready to receive them and who takes them out alive with his hands.

The ducks' necks were wrung, and they were put on the cart to the poulterers.

Decoying is alleged to have come from Holland in the reign of James I; 'decoy' is derived from the Dutch *eendenkooi*, meaning 'duck cage'. In their heyday there were two hundred decoys in Britain which slew half a million ducks a year. Most decoys were within thirty miles of the east coast. Possibly the biggest decoy was Fritton Broad, which was giant with at least twenty-three pipes. A variant on Defoe's fifth column ducks was to use the dog to entice the wild duck along a pipe; ducks are naturally curious and when they see a predator, such as a fox, they will keep it at a distance, but tend to follow it.

Decoys had an advantage over hunting ducks with shotguns, as the duck meat did not contain lead shot, thus a higher price could be charged for it.

In the mid-1880s there were forty-one decoys still in operation in England, and 145 which were no longer in use. Today there are only a few remaining duck decoys. These include Hale Duck Decoy in Cheshire, administered by Halton Borough Council, Boarstall Duck Decoy near Aylesbury in Buckinghamshire, owned and administered by the National Trust, a decoy in Abbotsbury Swannery, Dorset, and the Berkeley New Decoy at Slimbridge.

Completed in 1843, the Berkeley New Decoy covers 0.86 of an acre of open water, and includes four pipes. The best annual take in this decoy was in the year 1853–4 when 1,410 birds were caught, of which more than 1,300 were mallards. On 4 January 1854, 238 mallard, 7 wigeon, 1 pheasant, 1 hawk, 1 moorhen and 1 blackbird were caught in one pipe at the same time. This probably took place in a frost when the water in the pipe had been kept open by icebreaking. The catch figures were over a thousand again in 1861–2 and also in 1899–1900.

After that they fell off rapidly until the decoy became disused in 1929. The pipes were reconditioned in 1937 but the decoy was scarcely used and a considerable amount of repair was again necessary in 1946 before it could be put into working order for the Wildfowl Trust. Of course, the trust does not use the decoy to catch ducks for the plate but for ringing with small individually numbered metal or plastic bands for identification.

*

Why did I drop bird-watching? Adolescence. I realized the social awkwardness of binoculars, of hiding in bushes.

So, I stopped organized bird-watching, carrying binoculars around my neck like some umbilical extension.

I never stopped noting the birds, though, as my wife and children will confirm. 'Oh look, there's a xxxx,' I say to groans of three. One memorable occasion I even persuaded the entire family to Slimbridge to watch the evening flight of Bewick's swans land in the pond in front of Scott's old studio.

But, essentially, by fifteen my life as a young gosling was over.

The Farm Pond, Herefordshire
18 FEBRUARY: The sky the weird white of boiled fish eye. No leaves on the alder; the reeds of last year dissolved to tarry, slimy matting; the bramble an untidy ball of wire; from on high the monody of buzzard; the face of the pond, sheet-metal.

The grass cropped, the pointy sedge (quite energetically evergreen for a grass) cropped. This is how a pond in the meadow should be in February. Stripped back, to the bare minimum. So today is the day I move the sheep out.

Let the meadow grass and pondside plants grow. Let the new season begin.

The Mill Pond, Argenton, Western France
19 FEBRUARY: Rooks in the poplars; in the swatch of paint colours, the sky is turned to elephant grey.

It has rained for days, and trees which once lived on land now emerge from waste water of the end of time.

On the Argenton, a little grebe in winter plumage dives and swims breathtaking distances underwater, thus 'diverdapper' and 'dabchick' are among the bird's common names. About the little grebe the Elizabethan poet Michael Drayton versed: 'Now up, now down again that hard it is to prove / Whether under water most it liveth, or above . . .'

In six months she is the only bird to have espied me in my office eyrie at the top of the Mill House. We play a fun little game. She dives under the running brown water, and I plot the square where she will rise; a sort of ornithological take on the game 'battleship'.

Every time I fail to guess where she will cork up. She does twenty yards underwater, she manoeuvres between the islands, although she must be blind down there in the flood-murk. On each surfacing, she looks me in the eye and shakes her head. For ten minutes we vie with each other, until she tires of my miscalculations, and lets the current bear her triumphantly away.

The next day, again in my high place in the house; a green woodpecker begins hammering a hole in the willow below the pond, the bird's tail both fulcrum and support. We are at exactly the same level. Occasionally, the bird leans back, looks at the job, mimicking a builder. The yaffle starts at 10.16am, and finishes at 10.59am, when disturbed by me taking Snoopy, our Jack Russell, for a walk.

As is our French habit, we pay our respects to the pond, check in, get the news. There is a dead toad on the bank, its legs ripped off.

This first pilgrim to the ancestral spawning pond has been intercepted by a stoat, on the very edge of sanctuary.

In the afternoon, late and rooky, the village lights flaming through the brusque trees, a tawny owl plummets towards the pond, and takes another toad.

This is the pond as nature's meat platter.

THE POND IN SPRING

The Mill Pond, Argenton, Western France
13 MARCH: The hanging willow branches are the
gold, braided hair of a maiden; on the alders, shoals of
snaky, rust-hued catkins. Here, in France, at least,
winter is melting away into spring. Around the top of
the pond, the shoots of yellow iris are pushing through
in six-inch blades, a water-bed of daggers; every day
they 'weaponize' towards the swords of their maturity.

It is misty; and mist always changes the rules of
the world, so in some misplaced spirit of investiga-
tion, I reach into the water, and break off a gnarled
rope of iris tuber, about eight inches of it. Inside the
tuber, there is pink flesh; I lick it. The effect is a mace-
attack to the mouth; my throat is sore for more than
a day.

Yellow iris is yellow flag (on account of its large
petals), is Jacob's sword, is *segg*, from the Anglo-Saxon
for 'sword'. The knifeness of the young leaves means
that the water-margin plant is occasionally nominated
as the origin of the 'fleur-de-lys' of heraldry.

My presence notwithstanding, a pair of mallards
are engaged in the violent sex that prompted Chaucer
to exclaim of the breed, 'Thy kind is of so lowe a wre-
chednesse, / That what love is, thou canst nat see ne
gesse.' I have written that a definition of the pond is
that it is waveless. Well, it is wavy now.

91

I Love the Little Pond to Mark at Spring

I love the little pond to mark at spring
When frogs & toads are croaking round its brink
When blackbirds yellow bills gin first to sing
& green woodpecker rotten trees to clink
I love to see the cattle muse & drink
& water crinkle to the rude March wind
While two ash dotterels flourish on its brink
Bearing key bunches children run to find
& water buttercups they're forced to leave behind

John Clare

The Farm Pond, Herefordshire

20 March: The 'common toad' is anything but; it is a priority species under the UK's Biodiversity Action Plan.

As well as pesticides and habitat destruction taking their toll, toads are killed in their thousands every March by cars as the amphibians waddle across the road to their ancestral breeding pools. Spring is the only season when toads live in water. Orwell considered toads mating the sure sign that spring was acumen-in; writing in 'Some Thoughts on the Common Toad':

Before the swallow, before the daffodil, and not much later than the snowdrop, the common toad salutes the coming of spring after his own fashion, which is to emerge from a hole in the ground, where he has lain buried since the previous autumn, and crawl as rapidly as possible towards the nearest suitable patch of water . . .

At this period, after his long fast, the toad has a very spiritual look, like a strict Anglo-Catholic towards the end of Lent. His movements are languid but purposeful, his body is shrunken, and by contrast his eyes look abnormally large. This allows one to notice, what one might not at another time, that a toad has about the most beautiful eye of any living creature. It is like gold, or more exactly it is like the golden-coloured semi-precious stone which one sometimes sees in signet-rings, and which I think is called a chrysoberyl . . .

But I am aware that many people do not like reptiles or amphibians, and I am not suggesting that in order to enjoy the spring you have to take an interest in toads. There are also the crocus, the missel-thrush, the cuckoo, the blackthorn, etc. The point is that the pleasures of spring are available to everybody, and cost nothing.

For a few days after getting into the water the toad concentrates on building up his strength by eating small insects. He swells to his normal size again, and enters a phase of intense sexual ecstasy.

Sometimes, one comes upon heaps of ten or twenty toads rolling over and over in the water, one clinging to another without distinction of sex. By degrees, however, they sort themselves out into coupling couples. You can now distinguish males from females, because the male is smaller, darker and sits on top, with his arms clasped round the female's neck. After a day or two the spawn is laid in long strings which wind themselves in and out of the reeds and soon become invisible. A few more weeks, and the water is alive with masses of tiny tadpoles which rapidly grow larger, sprout hind legs, then forelegs, then shed their tails: and finally, about the middle of the summer, the new generation of toads, perfect thumbnails of the adult, crawl out of the water to begin the cycle anew.

How to tell the difference between the common frog and the common toad? When disturbed, frogs tend to hop and toads tend to crawl. Frogs have smooth skin, toads have warty skin. Frogs have black patches behind the eye, toads have large lumps (parotoid glands) behind the eye.

On the way home, driving along the lane in the dark, we encounter the toad parade: they cross the lane, which is damp-wet, rather than soaking-wet, in starts and stops. Stationary, they are white lumps, almost leaves; the car headlights hit their pulsating throats just before the tyres hit their bodies. We weave, we slow to a sympathetic toad-speed crawl but it is impossible to save them all.

Getting out of the car, I climb the stock fence and follow them down the field to the pond. There are hundreds of them. One by one they reach the edge of the pond, and wade in. Some come singly, some come mounted, the male on the back of the female like a fat Gillray jockey. They come this night, as they will come every night for a fortnight, all tramping with single-minded purpose, towards the pool where they themselves were born, and their ancestors before them.

The males in the pond's shallows are mating with characteristic vigour and utter lack of discrimination; in the torchlight I can see males copulating with males; one toad is humping a stone. My grandfather once encountered a toad 'loving' (he was always delicate in his language) a fish.

On reaching the water the males begin their singing, but the stars are out, so the frost chills their ardour. By 2am, the toads have hushed; by 4am there is the skin of ice over the pond, and the toads have gone to stone, and sunk, as stones do, into the bottom mud.

The softening rays of the sun re-kindle the mating process. Some late males arrive in broad daylight; such is the desperation, the urge, and they climb hurriedly on to the back of unattached females, encircling their chest with strong arms in a vice that is not relaxed until she has deposited her eggs. The female is much larger than the male of the species; the sexual grasping is technically 'amplexus', and is disconcertingly human. A hug – almost. The male's grip is improved by nuptial pads, spiny patches on the 'fingers' of the forefeet; 'hands', occasioned by seasonal hormones. Male frogs have the same accoutrement.

One male, lovelorn, tries to climb on the back of a copulating couple in the shallow waters, but the male strikes out with his long hind legs.

Another bachelor climbs on to a female, already occupied. Burdened by the weight, the female sinks; one male releases his grip, to return to the pond's surface, jowl-mouthed, lascivious.

During the breeding season, the skin of the male tends towards a greenish colour, a smoother texture. He has a release call for use when grasped by another male. The day is full of this sound, a disconcertingly human *ow, ow.*

*

24 MARCH: Francesca Greenoak's *All the Birds of the Air* on moorhens: 'There are several ritual movements to be carried out during courtship and the matt-black bodies are put into expressively sexual shapes.'

I concur: Skitty and her mate are on a flat platform made from last year's reeds performing a Berlin cabaret *c.* 1930.

At night a smooth newt waddles towards the pond, having emerged from its hibernaculum, the pile of logs for the fire that never quite made it to the house for burning.

At the water's edge, the sleep-slow creature pauses, as if trying to recall whether it is a land animal entering water, or a water animal entering land. Such is the amphibian's eternal philosophical conundrum.

October to February is the main over-wintering hibernation period for the three species of British newt, and this occurs on land.

What is the key to the waking from hibernation? Some shudder in the earth (as Orwell proposed), the penetration of birdsong, the earth warming, the air temperature? Newts usually come out of hibernation when temperatures reach over 5°C. The temperature according to the thermometer on the barn wall is 6°C. But how did the solitary newt, buried under half a ton of logs, know this? The warming of the earth, I suppose, rather than ambient temperature.

In every rushy corner and puddle, there is a pair of pop-up mallards. A grey wagtail taking the undulating line through the air, humpy, switch-back, wavy-sea, across the pond. In the alder, the notes of the great tit screech like a rusty pit-wheel.

The next morning, Big Bird, now dressed with the rosy-red bill of the breeding season, is happily stabbing the toads. Through binoculars, I see he knifes them in the nape of the neck.

The Mill Pond, Argenton, Western France
26 March: The pond up the lane; giant carp basking slothfully.

On the approach to the water, the light strikes the pond peculiarly, so the surface seems solid. 'Glassy' might be a cliché, but then clichés are true.

By the time I head back towards the gîte, it is dusk and a stone curlew is Pan-piping somewhere unseen across the dry, stony fields. The bird, new to me, is properly a 'thick-knee' and no relation to the curlews on our old farm in the Black Mountains. It gains the honorific 'curlew' on account of its similarly eerie call.

This evening, another stone curlew takes up the

theme, then one close by, so that the three concert together, haunting and restrained, in a vast landscape.

27 MARCH: I have decided that this shall be the Age of Aquarium, and purchase a plastic version from Leclerc's Jardinerie in Bressuire. The woman next to me is filling a basket with ornaments to put in her aquarium. Skeletons, castle ruins, treasure chests. She holds up a purple and green luminous plastic ghost, looks at me, questioningly. 'De trop?' I shimmy my hand horizontally, in the European language of maybe/peut-être.

As well as the aquarium, I buy a natty plastic scoop net for €2.60; made by Zolux, Nanolife range. I then drive with Freda to the back of the Intermarché supermarket at Nueil-les-Aubiers, where there is a community green area along the narrow, tree-lined Scie river. And a pond where frogs have already spawned. The pond is an ox-bow pond, a remnant of the river's previous course, filled with rain water, and a stream so small it barely qualifies the description. In my eagerness to get pond-dipping I have forgotten the key kit, wellingtons, so we hop from tussock to tussock to get to the pond. Human frogs.

Still Water

Up close, the pond is shallow and grassy, as temporary ponds typically are. The grass flops flat, enervated, presumably due to osmosis. There are maybe fifty balls of frogspawn; inert, and easy pickings, pale bright against the green. I take one small ball of speckled spawn in my hand; it sprawls, cold, uncontainable, like jellyfish. Juggling, I place this one glob in an aquarium, already two-thirds full of waiting pondy water.

Then the fun begins: we trawl the fronds with the net. Spring is the most exciting time of year for searching out pond life; the lengthening days trigger the growth of water plants and almost all pond animals feel the need to breed.

First sweep, a glistening newt on top of the detritus; placed on my hand, its feet itchy-tickly. (Childlike/childhood smile.) Although they look similar to lizards, they are quite, quite different; lizards are land creatures, and newts are true amphibians who split the year between water and land. Lizard skin is dry and scaly, whereas newt skin (which common newts shed once a week) is moist, velvety.

We replace the newt in the water.

Second sweep: a seething multitude of animals scraped up: raft spiders, boatmen, diving beetle, the red water mite (*Eylais sp.*), whirligig beetles, pond snails, squirmy larvae of insects.

Freda is a committed newt-hunter, always has been. She pushes the net through the pond with the savviness of the experienced pond-dipper; slowly, and into the sun.

Another newt caught; admired; replaced.

Next trawl of the net: a *Notonecta glauca*, a backswimmer. Scarab-shaped, the backswimmer hangs from the surface film, loitering, until it detects the vibrations of prey moving about; then it dives down at speed, using its oar-like back legs for propulsion, clutches the prey with its forelegs, pierces it with its rostrum. The beast can live for nearly a year.

From above, the lurking backswimmer has the appearance of a mercury blob due to its holding of air bubbles on its chest. The backswimmer is a textbook example of 'counter-shading': its light-coloured back, seen from below, blends into the water surface and sky. The rest of the body and its chest are darker and, when seen from above, blend with the bottom of the body of water in which it lives. Since the backswimmer is lighter than water, it rises to the surface after releasing its hold on the bottom vegetation. At the surface, it may either leap out of the water, fly, or resupply the oxygen tank.

The backswimmer goes in the aquarium. As does some pond mud, and pondweed, plus a clump of arrowhead (*Sagittaria sagittifolia*), so called in recognition of its distinctive arrow-shaped surface leaves; the underwater leaves are spiky. The combination of flat floaters with a large surface area (to catch as much sun as possible) and spiky underwater leaves that give minimal drag to water flow is typical of the pond plant. Aquanautic.

That night as the dusty motes settle in the aquarium, we watch en famille the *Jeeves and Wooster* episode where Gussie Fink-Nottle ('noted newt-fancier') shows Bertie how newts propose, imitating their mating dance; other Drone Club members mistake his description of amphibians' nuptials for the gen on a new jig, which they adjudicate 'better than the foxtrot any day'.

28 MARCH: The water in the aquarium, my personal and artificial pond, has almost cleared. A whirligig beetle is performing a comic turn, sparking round and round on the water's surface. The eyes are in two parts, enabling it to see up and down simultaneously. Like the backswimmer, the whirligig can trap air against its body for long submerged swims.

There is a lesser water boatman (*Corixa punctata*)

too, another aquatic herbivorous bug with oar-like hind legs, but one that swims the right way up. Disappointingly this is not the sub-species *Micronecta scholtzi*. Size for size, the 2mm *Micronecta scholtzi* is the loudest animal on earth. By stridulating with its penis down in the murky deep, it can bang out 99.2 decibels, and is quite audible to humans.

The surprise in the aquarium is a horse leech, sinister and elastic, stretching and contracting as it sidles along the plastic sides.

29 MARCH: In the relative heat of the house, the frogspawn has melted, the black dots released as bimbling tadpoles, which rest on the sides of the aquarium in a flurry of twenty musical notations.

30 MARCH: The Collins Pocket Guide *Freshwater Life* incants: 'Remember that in a confined space, freshwater invertebrates and fish tend to devour each other.'

My codicil: 'Remember that in a confined space, invertebrates tend to devour each other.' The backswimmer has eaten the leech.

31 MARCH: Some observations on pond life in an aquarium. Backswimmers actually swim on their front quite a lot, and will also eat carrion; raft spiders actually walk on water regardless of the availability of herbage rafts; tadpoles possess gold glittery bodies, and suck on algal sponge (they are vegetarian); pondweed floating in the water has the subtle geometric beauty of a double helix; the sinking and the rising of the diving bugs is almost rhythmic.

Outside, at the 'real' pond, the rooks are building their fuzzy nests, the frogs are calling at night, cuckoo during the day.

And if I were to return a hundred years from now would there still be starlings popping in and out of the hole in the willow, feeding their young? This, by the by, is the hole started by the green woodpecker. Always the chancer, the starling.

The wind gets up, and whips the willow tresses into a head-tossing frenzy.

> *At distance from the water's edge,*
> *On hanging sallow's farthest stretch,*
> *The moor-hen 'gins her nest of sedge,*
> *Safe from destroying school-boy's reach.*
> *Fen-sparrows chirp and fly to fetch*
> *The wither'd reed-down rustling nigh,*

And, by the sunny side the ditch,
Prepare their dwelling warm and dry.
From 'The Last of March', John Clare

The Farm Pond, Herefordshire

4 APRIL: Bright wind on the water, scurrying it; ducks moored, avian buoys. Somewhere beyond the hill, rooks honking like a fleet of taxis.

Last night's rain has created new ponds, which spangle with sun shards; a pied wagtail delights in one of the pools, stopping every half minute to peer at the water, head tilted, for some potential morsel, before fluttering off. John Clare caught their manner:

Little trotty wagtail, he went in the rain,
And tittering, tottering sideways he near got straight
   again
He stooped to get a worm, and look'd up to catch
   a fly
And then he flew away ere his feathers they were dry.

In the 'proper' pond (ponds are exercises in relativity) floats a dead female frog, face down; male frogs can embrace their partner so closely they crush them.

Frogs, unlike toads, are indiscriminate spawners; they will breed in any convenient water, as opposed to the ancestral pool. Ditches, temporary puddles are often used. The mass breeding of frogs at the pool is not determined by history, but by frogs being drawn by the siren sound of each other's voices.

The name 'frog' is a corruption of the Anglo-Saxon *frogga*. In the past it was believed that by inflicting pain on a frog one could, by supernatural forces, transfer the same pain to one's enemies. The frog was buried alive, so as to bring slow death to the chosen. The hapless frog was also worn around the neck in a silken bag as a talisman to ward off evil, a condition believed to be caused by the entry of Beelzebub into the body. Instead of entering the human, the devil – a lover of the ugly – entered the frog. Throughout the seventeenth century frogspawn was used as a poultice to stop bleeding. Tadpoles – from the Middle English *tadpolle*, or 'toad head' – were swallowed alive in the belief they cured gout. Adult frogs were placed in the mouth as contraceptive.

The French, of course, put them on the plate as food.

I do try. I go to Intermarché in Nueil, buy a packet of Gimbert's *Cuisses de Grenouilles*, 13 pieces for €8.19.

Into a frying pan I plop the frog bits, plus shredded vegetables, tamari, ginger, garlic, extra virgin olive oil – the full Jamie.

Unfortunately, I whack the gas too high and the spitting of the vegetables causes the frogs' legs to surreally hop around the pan.

Finally, I pick at the pale flesh; as all attest, frog meat is chickeny pleasant. But I cannot stomach eating more than a pale shred.

Carshalton Ponds, Surrey
15 APRIL: On the way to the ferry, via Carshalton Ponds, Surrey; I learned to drive a car around here, the VW Golf of Penny's mother, who was an utterly fearless instructor. She'd driven 3-ton lorries in the Second World War, a feminist feat that caught the eye of the *Daily Mirror*, along with her angular, Bacall good looks. They made a serial of her.

The Carshalton Ponds are English village ponds perfected; pub, swans, a green, mock-Tudor houses, mature trees. If no one is playing cricket or tolling church bells they should be.

The ponds in the centre of Carshalton date back aeons, probably explaining why there was an early Saxon settlement here. Originally a single body of water, the present duo of ponds was fashioned in the fifteenth century – with one for the public and one private for Stone Court. The ponds and the conservation area around them amount to 7.21 hectares, and they owe their picturesqueness to the Victorian aesthete and social reformer, John Ruskin.

In 1872 John Ruskin gained permission from the Lord of the Manor's Court to clean the existing ponds, cement the bottom, construct sluices and landscape the surrounding area, using 7 tons of Cumberland stone. Wooden palings were put up round the site, flowers and shrubs planted by the paths. He dedicated the site to the memory of his mother Margaret, and a stone there is engraved with the following inscription: 'In obedience to the Giver of Life, of the brooks and fruits that feed it, and the peace that ends it. May this well be kept sacred for the service of mews, flocks and flowers, and be by kindness called Margaret's Well. This pool was beautified and endowed by John Ruskin Esq., MA, LLD.'

In 1899 Carshalton Urban District Council took responsibility for 'Margaret's Pool', which is really a pond, but 'pool' perhaps sounds better to the serviette snobberies of the suburbs.

If extreme poverty conserves, so does extreme wealth. Carshalton is Surrey stockbroker land. The good burghers of Carshalton have no need of in-filling their gardens for extra housing, or catering for hoi polloi with a Maccy D by the water.

Carshalton Ponds were once used to power the local grain mill, and given its propinquity to the A232, an ancient east–west trackway below London, it was certainly also a 'horse pond' for the watering of equines, our native locomotion for almost two millennia. When a horse pond was constructed it was either lined with stone or 'puddling', a six-inch-thick layer of pounded clay and lime. The lime was to deter worms burrowing through and holing it, making it porous. Typically, post-medieval horse ponds were designed so that the horses, and their vehicle, could be driven in one end and out the other, to wet the wooden wheels and prevent their splitting. Constable's *The Hay Wain* shows the wheels of a horse and cart being thus hydrated.

Other utilitarian employments of the village pond: drinking water, flax-retting (soaking flax to release the inner fibre from the outer stalk), ice-making, dyeing, watercress-growing, droving (taking livestock to a waterhole on the way to market), watering traction engines.

Of course, the use of the village pond extended

beyond the utilitarian. Aside from skating and curling in winter, the pond was, once upon a Britannic time, the place of amusement and punishment for all seasons. A medieval jape was to tie an owl and a duck together on a short length of leather, and place the duck on the pond, the owl flapping above it. The petrified owl hooted, the duck dived underwater in fright, pulling the owl down with it. The 'sport' continued until the owl drowned. For centuries, supposedly transgressive women were ducked in ponds, which was penalty and spectacle in one. A 'ducking stool' featured a long oaken beam with a chair attached to one end. The beam was mounted to a seesaw, allowing a 'scold' – a 'nagging' woman – to be dunked repeatedly in the pond or river. The action would supposedly cool her 'heat'. In sentencing a woman the magistrates ordered the number of duckings she should have.

The CIA have been vilified for 'waterboarding', but water torture is nothing new. The earliest recorded use of the ducking stool is the beginning of the seventeenth century. Sometimes, however, the ducking stool was not a static fixture but was movable, a 'tumbrel', a chair on two wheels with two long shafts fixed to the axles. Before ducking, the tumbrel was trundled through the streets, the supposed termagant being pelted on her bare buttocks. On reaching the pond, the tumbrel was pushed into the water, the shafts released, thus tipping the chair up backwards, to dunk the 'harridan'. The last recorded ducking cases are those of Jenny Pipes, 'a notorious scold' (1809), and Sarah Leeke (1817), both of Leominster. In the last case the water in the pond was so low that the victim was merely wheeled round the town in the chair. The Priory Church in Leominster, Herefordshire, possesses the very ducking stool used to punish Pipes and Leeke.

In 1615 Samuel Pepys recorded in his papers a ballad about this method of controlling women, called 'The Cucking of a Scould':

> *A Wedded wife there was*
> *I wis of yeeres but yong,*
> *But if you thinke she wanted wit*
> *Ile sweare she lackt no tongue.*
> *Just seventeene yeeres of age*
> *This woman was no more,*
> *Yet she could scold with anyone*
> *From twenty to threescore.*
> *The cucking of a Scould,*

Still Water

The cucking of a Scould,
Which if you will but stay to heare,
The cucking of a Scoulde . . .

Then was the Scould her selfe.
In a wheele-barrow brought.
Stripped naked to the smocke,
As in that case she ought:
Neats tongues about her necke
Were hung in open show;
And thus unto the cucking stoole
This famous Scould did goe.
           The cucking, etc.

Then fast within the chaire
She was most finely bound,
Which made her scold excessively,
And said she should be drown'd.
But every time that she
Was in the water dipt,
The drums and trumpets sounded, brave
For joy the people skipt.
           The cucking, etc.

Upon which words, I wot,
They duckt her straight againe

A dozen times ore head and eares:
Yet she would not refraine,
But still revil'd them all.
Then to't againe they goe,
Till she at last held up her hands,
Saying, Ile no more doe so.
    The cucking, etc.

The village pond was also employed in witch-finding, being the venue of the laughably titled 'swimming test'. The suspect was stripped naked and then tied up – the right thumb to the left big toe and vice versa; in this position she was then secured by ropes and thrown into the pond. If she sank and drowned, she was innocent and would go to heaven; if she floated, she would be tried as a witch.

Women generally floated; it is a matter of the ratio of body fat to muscle.

This is the pond as place of murder. Paradise lost.

The official use of 'swimming' in English law dates back to King Athelstan (927–39), where trial by water, termed 'indicium aquae', was a general test for all crimes. It ceased to be an official law in 1219 under Henry III's reforms. For the next six hundred years it was popularly, but unofficially, supposed to be an infallible test of the guilt of witches and those suspected of subscribing to the black arts. The practice had a theological endorsement from no less than James I of England in his *Daemonologie* (1597), where he declared that 'God hath appointed . . . that the water shall refuse to receive them in her bosome, that have shaken off them the sacred Water of Baptisme, and wilfully refused the benefite thereof'. Water rejected servants of the devil and if a suspected person floated and refused to sink when placed in water it was proof of guilt. 'Swimming' was a reverse baptism.

Swimming witches was a favourite of Matthew Hopkins, a 'Witch Finder General' during the English Civil War. It has long been held that Hopkins was himself accused of being a witch, subjected to his own test of being bound and thrown into water and hanged after he was found to float. His ghost is said to haunt Mistley Pond in Essex.

Disappointingly for divine justice, Hopkins in truth died after an illness, likely tuberculosis.

Reports of 'swimming' witches occur quite frequently in the following decades, almost always carried out at the local level, and increasingly with the disapproval of the authorities. One of the last recorded cases in

Britain was in 1751, when seventy-one-year-old Ruth Osborne of Tring in Hertfordshire was suspected of witchcraft, grabbed by a large mob, tied up, beaten and thrown in a local pond. She was fished out again and laid on the bank, but died a few minutes later. The ringleader of the mob, Thomas Colley, was accused of wilful murder, arrested, tried and executed. His body was then hung in chains from a gibbet in Tring.

The Mill Pond, Argenton, Western France
15 APRIL. On the ferry to France, looking at the Channel I see that the sea is too brutal, energetic, fathomless for comfortable contemplation. The sea is not restful. You need a deep, forgotten pond for daydreaming.

At night in Argenton I walk up the lane; people regard night-walkers as criminals, robbers or rapists, so I take the dog. Everyone trusts a dog-walker. On the *étang* the Daubenton's bats circle in their tens. It's half-light at eleven o'clock. France is an hour ahead. From the *étang* comes the Dalek din of croaking frogs.

When I get back to the Mill House, the frogs have started up there too.

Later, about midnight, a line I never thought I would write: 'The frogs are so noisy I can't get to sleep.' *Rasp. Rasp.*

Whirligig beetles can trap air against their body for long swims. The species is a strong flyer too; when I take the lid off the aquarium on the landing (the Goldilocks place: not too hot, not too cold), they flee. Tadpoles defecate to prodigious degree, and despite the oxygenating pond plants I see a slight change in water colour.

I feel the mini-pond experiment coming to an end.

17 APRIL: Late evening, mackerel sky: tens of frogs, most of them in the throes of sexual congress, peer at me from just above the water's surface. As I stand there, more pop up – the males emitting a low-pitched call to attract the females; the frog collective's chorus is the mewing of a giant cat.

Some of these frogs will have been coming to the mill pond for years; eight years of life is unremarkable for *Rana temporaria,* if they can survive the predators. A big if.

Only once before have I seen a larger gathering of frogs, and that was years ago on a water-logged Herefordshire meadow. In one small grassy pool hundreds of frogs congregated, and the air reverberated with their singing, the sound of which I can only compare to a steel works in action. The tiny pool dried up quickly in spring sunshine, and none of the spawn hatched. Although the area flooded for years afterwards, the frogs never came back.

Gently, I tip away the contents of the aquarium into the mill pond. They swim free. Pond life to pond life.

In that twelve inches by eight inches by eight inches of aquarium was an entire universe.

Or even universes. I put a millilitre – no more than that – of aquarium fluid under the microscope. In this single, apparently clear, drop are kaleidoscope millions of microbes: green algae, red flailing rotifers, white protozoa.

When Antonie van Leeuwenhoek, a Dutch merchant and part-time janitor, turned his newly invented microscope on a drop of water he called his daughter, 'See what I see, Maria!'

They saw a new universe. In that moment Van Leeuwenhock and his daughter discovered more species new to science than all of the terrestrial explorers of the previous century.

*

STILL WATER

Tupsley Quarry, Hereford

19 APRIL: Like Georgie Bowling I am going back to see the splashing ponds of my childhood, where I roamed with my *Observer's Book of Pond Life*, a DIY *Blue Peter* net made from nylon tights, wire coat hangers and bamboo poles; the 'collection receptacle' was a Robertson's jam jar with a string handle or, best of all, one of those large Polish gherkin jars – also with a string handle – bought only at Christmas. (Later, I upgraded to a proper fishing net from the gift shop in Borth, and a plastic bucket shaped like a square four-turreted castle.)

The Quarry Ponds were the outer edge of the area I was allowed to roam, two miles from home; there was an invisible, acknowledged border to my free-ranging: Ledbury Road. Beyond that was Whittern Way, a sprawling redbrick estate where kids who rode Choppers lived. (Read 'rough boys'.)

They *were* quite rough, some of them. Among my babysitters were Mr and Mrs Abbott, who worked for my father, and lived off Whittern Way. Once, my father went into their house, leaving me to saunter to the newsagents on the square to buy the *Rover* comic (featuring Alf Tupper, the Tough of the Track. I was a runner and Alf was another boyhood hero). In front of the shop was a sunken playground, containing nothing more than a large concrete pipe (sewage version) to play in as 'tunnel', and some swings, already vandal-ized, the seats burnt. Outside the newsagents, a rat-faced urchin in a parka jumped me; undersized but accustomed to violence. (I may have exaggerated the rodency; ten years later Martin O'Neill and I came face-to-face at a Young Farmers disco, buying the same girl a drink; such are the gin-traps of time.) I was in my M&S anorak and brown sandals; an invitation to a kicking. I'd have kicked me.

'Fight! Fight!' A circle, of adults as well as kids, formed around us in the time-honoured manner. Someone went to get my father, who duly appeared as we tugged and punched.

A woman asked my father to stop the pugilism. He looked at us both, and said 'No', and went away again.

I have never loved my father more than in that moment.

I think I won – at least it was a draw – although nothing is clear in the memory until we got home to Hampton Park and my father handed me a bloody steak from the fridge to put on my black eye. (His quo-tidian supper was Steak Diane.)

Anyway, this April day I have decided to imitate Georgie Bowling. I enter the Quarry on foot via the thin, fenced alleyway from Church Road, which dis-gorges me into the playing field area. Where did those horse chestnut trees along the paths come from? It must be well over thirty years since I was last here. Can trees really mature that quickly?

Any inclination I had to grumble à la George Bowling ('When I was a boy all this was . . .') is checked immediately by a large coloured sign announcing 'Quarry Park', and a picture of an idyllic pond. It transpires that the Quarry is now a local nature reserve:

> Tupsley Quarry is an important site locally for amphibians such as frogs and newts. It has seasonal ponds that appear in the winter months when the weather is wet and are usually around long enough for them to lay their eggs (spawn).

Although most of the park is goal-posted playing fields and paved paths, a third is still wild, ponds and scrub.

The ponds of my childhood are almost exactly where I left them, three of them, behind a long, tall bank formed from the spoil from clay pits, now as then covered in dense, ivied thicket, and tunnelled by children convinced they are Allan Quatermain/Indiana Jones (delete as per your demographic).

The sun is shining; someone on the cliffette above the ponds is lawnmowing; three young women in 'jeggings' are trundling kids in pushchairs around the playing fields, exclaiming; a man is shouting 'fetch' to a golden retriever as he slings a rubber ball; the sound of the suburbs.

To get a good look at the ponds I have to go off the beaten path, and clamber a mud track, shadowed and secret, over the bank. It's an up-and-down mud-slide.

The ponds are shallower, more overgrown than they were in my time, and riddled spiky with pussy willow. Trees growing in water are a sign of approaching 'pond death' through choking, but The End is still some way off for the Quarry Ponds. A mallard appears from behind one tree, and tacks around the brown-water obstacle course of the others. Then, seemingly instructed by the LNR board, a great crested newt surfaces, as if to prove the ponds' claim of being an important site for amphibians. Indeed, it is where conservation groups do their 'great crested newt training'.

Maybe the great crested newt is a messenger from my past. There comes into my mind one of the most overwhelming, precise, scenes from my past ever; a great crested newt I caught here, and placed in the red sandcastle bucket, full of water, so it glowed in its own private sunrise. In every way it was a miniature dinosaur, lemon body and ruffled back ('crested'), down to its rubbery tail. But it was not rubber, or an image on TV.

How do we make children interested in nature? Words and pictures are not enough. We need the nature for them to be interested in. A great crested newt in my red plastic-castle bucket did it for me.

The ponds here, when I was about nine or ten, are the only place and time I have seen great crested newt courtship; it was night, it was spring, the shallow side of the middle pond, and the male swam around the female and nudged her; then did a handstand whilst waggling his tail (that brought a cheer from the four of us) in the spotlight of the Ever Ready pocket torch.

If the female great crested newt is receptive to such dancefloor moves, the male transfers a spermatophore, a gelatinous jelly cone containing semen, picked up by the cloacal lips of the female.

In total, female great crested newts will lay between two hundred and three hundred eggs in the water, usually two or three at a time. When laying the eggs, the female uses her hind legs to individually wrap the white 5mm-long eggs in leaves found in the pond or overhanging vegetation. The host vegetation is easily identified by a characteristic 'concertina' appearance. Wrapping the eggs protects them from UV damage and predation.

After four weeks the eggs hatch as tadpoles and after a further three to four months, they develop into juveniles – called 'efts' – capable of leaving the water. The efts leave the water to hunt on land, and share the same carnivorous diet of slugs, worms and insects as the landlubber adult newts. At this time, the young newts will spend anywhere between one and three years on land until they become sexually mature.

Great crested newts possess low-down, amphibian cunning. They can literally 'sniff out' ponds that are fish-free havens with a life-affirming neutral pH. The best underwater locations for the courtship dance are hotly contested, with some males mimicking females to lure rivals away and occupy the site themselves.

I'm thinking this, as I scramble back out from the ponds on to the civilized footpath, dishevelled by the scratching-clutching limbs of the hawthorn trees, coat askew, hair wild. I debouch into the path of a woman walking a dog. In my hands I have a notebook and pen; 'I'm making notes on ponds,' I say, c. 1972 *Monty Python.*

The great crested newt is a large-scale amphibian (hence the 'great'), with adult females reaching 17cm, and both sexes possessing the lumpy skin that accounts for its colloquial name of 'warty newt'. The underside is orange with black spots; on females, the orange colouration continues along the underside of the tail. The jagged crest is peculiar to the male in the breeding season.

Great crested newts and their habitats are fully protected by the Conservation of Habitats and Species Regulations 2010 and partially protected under the Wildlife and Countryside Act 1981 (as amended). It is an offence to kill, capture or disturb them, or to damage or destroy their ponds. A licence is required if disturbance of great crested newts or damage to their habitat is likely to occur as a result of a development or other works.

The legal protection is due to rapid population decline, largely because of a reduction in suitable habitat due to pollution, and the filling-in of ponds by agriculture and urban development.

The tale of the great crested newt is a motif for everything that has gone wrong for freshwater wildlife in Britain. Tupsley Quarry LNR is a way to a different future.

The extraction of clay for bricks was the original use of the Quarry, and there were brickyards on site until 1937. The cherry-red bricks built Hereford's Odeon cinema and Ritz cinema; later the site was an

American Army camp in the Second World War; at D-Day they all left, and German PoWs were housed there. In 1946, locals forcibly took over the camp due to a shortage of accommodation.

Now the Quarry is local playground and local nature reserve. What better fate could there be for a weal of industry?

Britain possesses three species of newt – aside from the great crested newt, there is the common or smooth newt, and the palmate newt. The latter is the likeliest visitor to the garden ponds of Britain.

Common newts are olive green or pale brown with a bright orange, black-spotted underside. In the breeding season the male sports a wavy crest from head to tail. They may reach 11cm in length, with the male slightly larger than the female. Their average lifespan is six years.

A smooth newt can be told apart from a palmate newt, which is a similar species, by the presence of dark spots on the underside of the throat and the absence of a black mask around the eyes.

The Farm Pond, Herefordshire
22 April: Toads' eggs are easily identified. Instead of the dense clumps of spawn laid by frogs, toadspawn is

laid out in strings containing a double or triple row of eggs (a single row of eggs would indicate the rare natterjack toad). A toad 'chain' in the pond measures six feet when I untangle it (after half a careful hour) from the weeds and submerged branches; the female moves around as she lays the spawn. The chain might contain three thousand eggs.

It is a literal food chain; heron, buzzard, moorhen, fox have dined on it.

24 April: The brooklime is in flower, bewitching blue, going on mauve, hence 'water purple' locally, and so overlooked that one almost feels sorry for it. It likes to grow in slow freshwater brooks and at pond edges, but will make do with any damp place. The succulent, oblong leaves are robust and fleshy, bitterer than those of watercress with which it often shares a habitat. Usually sprawling, never more than 30cm in height, a member of the speedwell family, brooklime was well known in historical times as an antiscorbutic. The Stuart herbalist Gerard advised:

Take the juice of Brook-lime, Watercresses and Scurvy-grass, each half a pint; of the juice of Oranges, four ounces; fine sugar, 2lbs; make a syrup over a gentle fire. Take one spoonful in your Beer every time you drink to cure the Scurvy.

Today the bees make a direct line for it; then again bees are creatures of direct flight, unlike the butterflies, which seem to waft wherever the fancy takes them. All around the pond the reeds, the yellow iris, the sedge and the grass are greening and growing. The alder are in leaf, giving the pond dark eyelids.

1 MAY: The mudbuilders, martens and blackbirds, at their brick factory, the pond edge.

A sparrow is bathing in the shallows amid indolent, sybaritic sunning tadpoles. The moorhen is nesting in the briar: a shallow bowl of leaves and stems, seven eggs, the first of her two or three broods between now and August. She's in a thorn fortress; to see the eggs I stick a hand mirror on a long stick, and carefully ease it in through the entrance tunnel; the slinkiness of moorhens is useful for activities other than courtship. There are eight eggs, which are white and lightly Pollocked with brown.

Pond: late afternoon, the watery romance of it all; the water is clear (it has not rained for a week, and rain muddies as well as refreshes) and for once one can see through to the multitudinous life in such a small space; the diving dytiscid beetles, the darting larvae, the tumbling red mites.

Cuckoo flowers, also known as lady's smock, are in bloom along the bank, and the rhubarby burdock, lover of warmth and wet, reaches up to my knee. Gerard mentions that the young burdock stalks, peeled and eaten raw with salt and pepper, or boiled in meat broth, are pleasant eating, and stir up lust. The greater distinction of *Arctium lappa*, points out Geoffrey Grigson in *The Englishman's Flora*, is that it 'must have been more painted than any other plant – by Claude Lorraine, and by his landscape followers in England. George Stubbs, for example, is the Master of Burdock – burdock in the foreground, an oak sloping across the middle distance, and a racehorse in between.'

The pond flowers of spring make a pretty necklace. Orange tip butterflies feed on the lady's smock (*Cardamine pratensis*).

The Mill Pond, Argenton, Western France
3 MAY: Rings of ripples from nosing fish, rings of ripples from falling willow leaves. Rings from below, rings from above. But intersected by the straight lines

of light wind on water; noughts and crosses on the surface of a pond. Such beautiful geometry. Mobiles of midges hanging in the air; poplars stutter-shining in the breezy sun, disguising the male golden oriole, brightest of birds, as it sings its tropical cry of *we-ola*. Penny whistles to it; the bird whistles back.

Light chatter of starling chicks in the willow. A wood pigeon, disturbed, plummets out of the tree, then rises; the plunge-low to sweep-high is an escape plan they share with blackbirds.

Though a month advanced, the pondside flowers here are cossetingly similar to those at home: water forget-me-not (*Myosotis palustris*), hairy bittercress, watermint, red clover, buttercups. The watermint is already six inches high. The flowers are tourist-crowded by butterflies and bees and bumblebees. On the lane, cow parsley, musty and herby, rims the tarmac.

At night: nightingale song and star light; both reflect off the silver surface of the pond.

4 MAY: On the pond, a white dusting of pollen from the trees and flowers; in the heat, the bottom of the pond begins the belching season, with methane releases from decaying organic matter. Sometimes the belches are atomic, mushroom clouds of mud.

Some willow leaves drop early, to lie on the water like little gondolas. Ponds, they play pleasant havoc with the imagination.

The Farm Pond, Herefordshire
6 MAY: I did something today I have not done since I was a child: I watched a leaf burst its casing; and it occurred to me that we only 'see' a leaf as bud and as fully mature. We miss out the stages of growth; all leaves at first are small and undersized, even those of the horse chestnut, but in a little while they swell and grow exactly like the wings of a newly hatched butterfly.

High in the sky a heron came over, scanning for watery places. Other birds search for food; the heron looks for a place where prey might be. No bird knows topography like the surveilling heron.

Hawking for heron (particularly with peregrines) was popular in the Middle Ages, though the bird's tendency to avoid conflict by ascending gave it the stigma of cowardice. Edward III was presented with a dish of roast heron; cowardly bird for a cowardly king who refused to invade France. (The jibe worked, perhaps too well. Edward then led the English into the Hundred Years War.)

The heron was a regular dish on the medieval

banqueting table. As the property of the crown, heavy fines were levied on anyone caught poaching the bird, while in Scotland the penalty was amputation of the right hand.

Thornbury Fishponds, Gloucestershire
7 MAY:

> 'Streets, trains, cars, aeroplanes, houses, houses, and yet more houses . . . where will it end?'
>> BB, *Letters from Compton Deverell*, 1950

Returning from visiting Penny's aunt in Somerset, we detour to visit Thornbury's medieval fishponds, stopping first to get petrol at Gordano services; as I get out of the car I think the white splodges across the forestation tarmac are lichen; they are dried chewing gum. False lichen.

I'm already in an evil mood due to the amount of traffic on the M5. Is the entirety of Britain on the move? Bowling-Orwell at his uttermost dystopian failed to foresee the cars choking Britain to death. Doesn't it make you puke, what we are doing to Britain?

All I know about the location of Thornbury's fishponds is that they were part of the local castle, and are a scheduled monument 'under the Ancient Monuments and Archaeological Areas Act 1979 as amended as it appears to the Secretary of State to be of national importance'.

The Thornbury fishponds are scheduled for the following principal reasons:

> Survival – the fishponds survive particularly well with many of the pools retaining water. They are an especially intact group retaining a range of features. Period – fishponds are very representative of large scale animal husbandry during the medieval and post-medieval period. As such they have considerable historic interest. Potential – the likelihood of water-logged deposits means that the site has the potential to retain an especially good range of artefactual and environmental evidence. Group Value – the historic interest of the fishponds is considerably enhanced by their association with Thornbury Castle (listed Grade I).

Thornbury Castle is now a plush hotel, so we drive there, in our seeking of the ponds. The young receptionist in the panelled hall looks blank behind her glasses at our odd request, but has the grace to ask someone else. From a backroom appears a woman in her carefully preserved sixties, wearing a Russian fur hat.

She is politeness personified. 'It's ages since I've been there,' she explains, 'but go down past the school, and turn left into a small lane, into the housing estate. And they are building more houses there.'

She looks almost tearful. 'So many houses in Thornbury now. We've had seven hundred, and they are building another two and a half thousand. It's only a small town. It's difficult to park, especially if you have an elderly husband, like I have.'

Her directions are more or less accurate; we take a lane into a housing estate, which stretches this way and that way, but head downhill. Leaning out of the car window, I ask a woman delivering community action leaflets the whereabouts of the ponds. 'I haven't been there for years, but walk down there to the woods.'

We get out of the Saab and the ponds turn out to be a mere hundred yards away, at the end of a tarmac path, through a kissing gate.

Clearly nobody goes there. The public ponds are almost entirely surrounded by a builder's wire fence: 'Danger Deep Water. Keep Out: David Wilson Homes Where Quality Lives.' A yellow Labrador dog adorns the sign. Like the Andrex one.

It's late afternoon; a chiffchaff is mindlessly mimicking the sound of cutting metal, a blackbird is singing vespers, or maybe a requiem, in an ancient oak, whose split trunk makes legs for an Ent. Mars looms in the sky.

According to the paperwork of English Heritage, the fishponds at Thornbury comprise a series of:

> eight interconnected pools of varying size and depth arranged in a tight group of three broadly rectangular ponds, aligned north-west to south-east, at right angles to either side of a rectangular central pool with an outlying pond to the south. The ponds, used for the breeding, raising and storing of freshwater fish, are fed by springs and a stream to the west and are interconnected by a series of leats and sluices to allow control of the water levels within them. The ponds range in size from the smallest at c. 8m by 6m, probably used for the breeding of fish, to the largest at c. 29m by 6m, with individual ponds being used for different species or sizes of fish.

In real life? The first pond is silted and mordant under the crowding alders and hazels; the ground around tangled by foot-clutching ivy. Floating on the expanse of duckweed is a Lucozade bottle, and a plastic plant pot. By another pond, smoke-blackened timber.

So this is Bowling returned to Binfield. The Thornbury fishponds were supposedly 'saved' by English Heritage, and at some stage in the recent past a boardwalk around the seven ponds was laid down. Now it is rotting, and torn up for vandal fires, primed with paint.

Even with the neglect and the fenny gloom it is possible to understand what a brilliant facility the fishponds were, connected by stone sluices, with fresh water running through them from the stream.

But this is Bowling returned to Lower Binfield. Trudging through to the downstream end of the ponds, I gaze through the wire at the new building site, with its flagpoles. (Why do construction companies fly flags? Imperial conquest, I presume. 'We claim this land in the name of David Wilson Homes.')

The Thornbury ponds have been saved in name only. They are throttled by housing, all scale, wonder and biodiversity gone.

In England, the first large-scale building of artificial fishponds was undertaken by the Romans, who brought their pisciculture with them when they invaded.

Writing around 37 BC, Varro provides the earliest account of fish farming. Many fishponds were located adjacent to villas, in seaside coves and inlets or in lagoons, where they could be fed by both salt- and freshwater. Varro is scathing about the vogue for fishponds, asserting that their true appeal lay in their ostentatious proof of wealth: 'For in the first place they are built at great cost, in the second place they are stocked at great cost, and in the third place they are kept up at great cost.' To prevent the fish from escaping, brass gratings were placed across the channels. Cicero, like Varro, has no patience for all this bling, calling aristocrats such as Lucullus and Hortensius *piscinarii* ('fish fanciers') and 'Tritons of the fish ponds'. The common people excavated *dulces*, or freshwater ponds, to farm fish for garum, a sauce consisting of fermented intestines and made to secret recipes, never divulged, the Colonel Sanders of the Ancient World.

The earliest known Roman fishpond in England is at Eccles, in Kent, and dates from AD 65–120. As in the motherland, fishponds tended to be attached to the major villas, and their intended un-naturalness, their proving of the Roman control over the natural world, was emphasized by their regularity of shape; they were either square or rectangular. A standard measurement was 30m x 10m.

After the Roman de-camping in AD 410, no fish-ponds existed in Britain, until the members of the Norman secular aristocracy constructed them to enhance their status. But the boom in fishponds came in the fourteenth century, and was partly due to more rigorous observance of the Christian calendar. Since the medieval church calendar contained both fish days and fasting days (when fish was the only permissible flesh), a quantity of fish was necessary. Banquets also required a reliable source of high-status fish such as pike and chub. (Pike, with its bones, is unpopular with modern mouths, as is the chub, another conspicuous piscine indicator of rank and wealth in the past.) Hence the construction of fishponds, for the breeding and rearing of fish, such as those at Thornbury, which were likely constructed by the Duke of Buckingham c. 1500.

Fishponds required a deal of skilled construction, being clay- or timber-lined and fitted with stone water channels. As the Middle Ages progressed, these ponds became more and more elaborate. Aside from their construction cost, they required labour-intensive maintenance in stocking, husbandry and cleaning. They were a significant investment. The only people who could afford them were the aristocracy and rich monasteries.

The poor ate little fish, just as they ate little meat. Eels, roach and dace were the fish for proles.

Medieval fishponds had a steady flow of water through them to prevent the water becoming stagnant. The water was also relatively shallow because the fish needed to be caught easily. By the same token, the banks were kept free of growth so a net could be dragged across the pond.

Bream was by far the most popular freshwater fish on the royal table in the thirteenth century, and it retained favour for nearly four hundred years among the well-to-do and the discerning. In *The Compleat Angler*, 1653, Izaak Walton noted:

The bream, being at a full growth, is a large and stately fish. He will breed in rivers and ponds, but loves best to live in ponds, and where, if he likes the water and air, he will grow not only to be very large, but as fat as a hog . . . The fish is long in growing, but breeds exceedingly in a water that pleases him; yea, in many ponds so fast as to over-storen them, and starve the other fish . . . But though some do not, yet the French esteem this fish highly, and to that end have this proverb, 'He that hath breams in his pond is able to bid his friend welcome.' And it is noted that the best part of a bream is his belly and head.

By the time Walton wrote the above, the carp, a native of the Danube, had become *the* pond fish of intensive farming. Piscatorial treatises, such as that written by John Taverner, *Certaine Experiments Concerning Fishe and Fruite*, 1600, give them pride of place above all other fish.

The reasons for the carp's introduction as a popular species are plain. The secularized fishing industry of England, in an age of increasing materialism, could not have failed to notice the opportunities awaiting it. Carp grow rapidly, and reach large weights. Taverner claims to have seen pond carp in 1600 that were thirty-three inches from between the eyes to the fork of the tail. Such a carp, even allowing for poor condition factors, could be expected to exceed twenty pounds weight.

Commercial production of carp enabled the middle classes to ape the plates of their betters. In *The Canterbury Tales*, Chaucer's Franklin has a fishpond. In Hampshire, early in 1537, one Thomas Wriothesley is recorded as being about to stock the former monastic ponds at Titchfield with them. The document tells us how they were to be transported to the site, as well as indicating that commercial dealing in fish was now acceptable amongst the rising classes of society:

The bailey of Gernsey and Mr Wells of Hampton were here at the same time. Viewed the ponds – four of them a mile in length. The bailey will give Wriothesley 500 carp to stock the ponds, Mr Huttuft providing the freight, Mr Mylls tubs, and Mr Wells conveyance of the carps, so that in 3 or 4 years' time he may sell £20 to £30 worth of them every year . . .

Fish were money, and fishponds became a favourite target for poachers. In 1590 a John Pyke was accused of having stolen 'many and great carpes' from the Bishop of Winchester's fishpond at Frensham, near Farnham, in Surrey. In 1538 it is recorded that the Gyffard family of Suffolk had been about to supply the King with carp for his pond at Comebury when misfortune overtook them:

About Christmas 2 years ago the King wrote to Thos Gyffard to provide carp and other fish for his manor of Comebury. He drew his pools and put the chosen fish into Theves pond, which was robbed in the night, before they could be taken to Comebury, by persons from Claydon. Raffe Gyffard stole 5 carps worth 5 crowns out of the same pond, and on the 6th of this August, the Warell's pond was robbed, and a cart was traced to Steple Claydon, where Raffe Gyffard lives and nets and other things were found there. Raffe told those sent by Thos Gyffard that he would fish his ponds before his face.

Peasant discontent often expressed itself by assaults on fishponds, which were seen as bastions of privilege. The pond of the past was the unlikely site of class war.

Fishponds remained popular in the post-medieval period with landowners, as well as fish-stealers. The antiquarian Roger North indicates that freshwater fish are still considered luxury items at the beginning of the eighteenth century:

But you may contrive to keep your Stock (of fish) within Compass; for you may enlarge the Expence of your House, and gratify your family and friends that visit you, with a Dish as acceptable as any you can purchase for Money; or you may oblige your friends and Neighbours, by making Presents of them, which, from Country-man to the King, is well taken . . . it is a positive Disgrace to appear covetous of them, rather more than of Venison, or any other thing; so

that Presents are not only expedient, but necessary
to be made by him that professeth a Mastery of fish.

The date when fishponds ceased to be managed as
productive sources of fish is not known; most likely,
this would have been in the eighteenth century, or
even the early nineteenth century. At some point
between c. 1800 and the 1950s tastes changed, and fish
declined in popularity as a table dish. Unless you lived
in Herefordshire, where Friday was boiled white fish
even in the 1970s.

In 1952 Richard Walker, a fisherman of near Messianic
status among those who wait waterside with a rod,
broke the angling record for carp, hauling a 44lb
(19.9kg) specimen from Redmire Pool, near Ross-on-
Wye in Herefordshire. Walker christened the 37-inch
fish 'Ravioli' after the cold tins of pasta that he and his
companions ate at Redmire. A London newspaper
renamed it 'Clarissa'. The name stuck, and Clarissa
went to live in London Zoo's aquarium. The same
three-acre pond has produced other leviathans. Chris
Yates caught a then record UK carp of 51.5lb (23.4kg)
at Redmire in 1980 on a split cane rod that Richard
Walker himself had made.

Redmire pond dates from the 1600s, but the carp

fishing interest began in 1934 when the then owners of the estate instructed a fish supplier by the name of Donald Leney to stock the pool with fifty small carp (5.5–8 inches) in a bid to combat the extensive weed in the pond.

Why does Redmire produce such big fish in a small pond? Something in the water, perhaps.

The Mill Pond, Argenton, Western France
12 MAY: Luminous sky, the hue of starched white shirt.

The nightingale in the alder at the pond's end: allows one to get close up, within ten feet. They sing by day, they sing by night.

The wren: low brown-moth blur across the water, issuing its mechanical alarm; dragonflies flying past on wings of faeries; in the water, boas of green algae; frogs along the bank, secreted in the grass.

Every yard I walk, a frog pops into the water.

How far can frogs jump? An experiment half an hour later, with a builder's tape measure. Re-walk the bank edge, measuring the distance of the frog long-jumps. Four feet two inches is the day's record.

On the lane to Bressuire: cow parsley jewelled with black and scarlet beetles, a golden oriole whistling, but again I fail to spot the bird; as well as the secrecy of shining leaves, it throws its voice (I am sure). Sometimes it is the only sound in the jungle.

Standing on the bridge over the Argenton, I watch the kingfisher fly through the span below me: a purposeful, speeding bolt of cobalt fired upriver.

The Kingfisher

It was the Rainbow gave thee birth,
And left thee all her lovely hues;
And, as her mother's name was Tears,
So runs it in my blood to choose
For haunts the lonely pools, and keep
In company with trees that weep.
Go you and, with such glorious hues,
Live with proud peacocks in green parks;
On lawns as smooth as shining glass,
Let every feather show its marks;
Get thee on boughs and clap thy wings
Before the windows of proud kings.
Nay, lovely Bird, thou art not vain;
Thou hast no proud, ambitious mind;
I also love a quiet place
That's green, away from all mankind;

A lonely pool, and let a tree
Sigh with her bosom over me.
W. H. Davies

Croft Castle, Herefordshire
18 MAY: Passing by the entrance to Croft Castle, a National Trust great house, on my way to Cumbria to give a talk (to the National Trust, as it happens, at Wordsworth House), I thought I'd drop in, and see the restored fishponds.

So I drive up the long track, which conveniently runs the edge of the wooded valley (which deserves a capital V, because that is the landscape feature's cross-sectional shape) where the fishponds are located. The sun is hard and fresh with springness; the sky is the blue of bird egg.

This is a wholly different piscatorial enterprise to Thornbury, simpler, almost primitive; a stream dammed in a gorge, to make five descending pools. Proper, rustic Herefordshire, where we only imitate moneyed culture.

What the fishponds of Croft Castle lack in the expensive engineering of the past, they make up for in contemporary glory. The open ponds sing with vitality; the water is seductive, smoked grey right through to the stony base.

In celebration, a trout leaps clear; the sun scorches off its scales, and the individual drops of water are brilliant, cut diamond.

The fishponds here also used to provide ice. In winter it was cut in blocks and stored for summer in the ice-house, a heavily insulated brick vault.

The wood anemones are in flower, as white as ice themselves. The trees echo with the scrapyard notes of the chiffchaff.

On to the Lake District. Sitting on the beach at Bassenthwaite, rippling waves with white crests, catching the sunlight. Water sprites. Tiny dancers.

Lake District, Cumbria

20 MAY: The weight in my backpack is heavy when we start the climb for Great Gable from Honister Pass car park. I'm carrying the duty of Remembrance.

This is about nine in the morning. The Lake District weather is glorious so the mountains are backlit by beauty. I'm accompanying 'The Fellowship of Hill, and Wind, and Sunshine', an amateur choir gathered together by National Trust worker Jessie Binns. To mark the occasion of the centenary of the end of the Great War the fellowship will sing on the Lake District peaks, in commemoration of the fallen.

But in celebration, too, of 'The Great Gift' – the thirteen peaks gifted to the National Trust in the aftermath of the Great War by Lord Leconfield and the Fell and Rock Climbing Club. The names of the peaks run like a creed through the mind of every lover of high places. There is Great Gable itself, then Scafell Pike, Lingmell, Broad Crag, Great End, Seathwaite Fell, Allen Crags, Glaramara, Kirk Fell, Green Gable, Base Brown, Brandreth and Grey Knotts.

The mountains were given in memory of the dead, and that was fair and fitting. The men of 1914–18 were nature boys, none more so than soldier-poet Lieutenant Noel Hodgson, who died on the first day of the Somme, aged twenty-three. Great Gable was his favourite place in the world. He had taken a 'Farewell Walk' of the peaks before leaving for France, with his friend Nowell Oxland, another fatality of that bloody war. Lines from the poem Hodgson wrote for the dead Oxland could be said about himself:

You were a lover of the hills, and had
From them some measure of their Roman strength . . .

Great Gable is well named; it is the same pyramid shape of the end of a house, but wrought Himalayan-large and in bleak grey stone. The summit of Great

Gable is 899m above sea level, and the views over the purple mountains and bright lakes are to die for which, in a sense, Hodgson did. Topping the mountain is a plaque bearing the name of twenty Fell and Rock Climbing Club members who lost their lives in the Great War.

Across the valley is the awe-inspiring Scafell Pike, 978m, the loftiest peak in England.

I look at the uncompromising boulders strewing the summit of Great Gable, and I do not think it far-fetched that the geology of place must have entered the flesh of men. Both Hodgson and Oxland died in action, leading their troops.

The choir assembled in a horseshoe, and sang the fellowship song from the top of the world, and the words took off on the breeze across the valleys. 'Upon this summit . . . for freedom and sunshine . . . our spirit shall endure.'

I thought about Hodgson, whose grave in France I have visited twice, in tribute, because he once inspired me to write a book and it was true and good.

I came off the mountain alone, because I wanted to enjoy the Great Gift in its magnitude: the right to 'wander lonely as a cloud' among the peaks and meres of the Lake District. The gift was not just for the dead, it was for the living too.

Descending Great Gable, there was just me. No other human in sight.

A whinchat perching on a boulder *tic-tacced* absent-mindedly. There were white alpine flowers in crevices. Some Herdwick sheep grazed safely.

And there was the upland pond; black water, cracked mud, no sign of life in its shallow water. Moorland pools are poor in terms of wildlife diversity. Few creatures like to live in an acid bath.

The mere only lacks a tree to be a passage in *Beowulf*:

Mysterious is the region they live in – of wolf-fells, wind-picked moors and treacherous fen paths; a torrent of water pours down dark cliffs and plunges into the earth, an underground flood. It is not far from here, in terms of miles, that the Mere lies, overcast with dark, crag-rooted trees that hang in groves hoary with frost.

The Farm Pond, Herefordshire
21 MAY: A smooth newt, head raised, mouth open, half in and half out of the water, rubs against a stone to roll back its neck skin, then alternately rubs the sides of its body to push the roll of skin towards its

149

tail; then steps out of the skin, as if stepping out of four wellingtons. Suddenly, it twists around and seizes the rolled skin, now down towards the tail, and tugs.

Hey presto! The newt reveals its new clothes.

Damselflies on the wing, the blue male's posterior gripping the top of the green female's neck (as Roger Deakin noted, the arrangement looks like 'aircraft refuelling in flight'); they cruise backwards and forwards over the water like this, the male showing off his trophy wife.

These slender flying insects – country people used to call them 'Devil's Darning Needles' – make up the scientific order *Odonata*, with damselflies being the sub-order *Zygoptera* and dragonflies the sub-order *Anisoptera*. With their brilliant colouration, they are flying jewels of the British natural world. Damselflies and dragonflies are closely related, but generally damselflies are smaller and always rest with their wings closed lengthways against their bodies; dragonflies rest with wings outspread; and wing venation is simpler in damselflies than dragonflies, so they are weaker fliers.

Largely unchanged in appearance since pre-historic times, both sub-orders share a mating method requiring

Karma Sutra dexterity. When interlocked, the two insects form a 'wheel' or 'heart' shape. Depending on the species, mating takes place on the wing, or perched on vegetation or the ground.

Dragonflies and damselflies are marvels of aeronautic engineering and among the most effective predators in the natural world. The angle and beat of their four wings can be controlled independently, which allows dragonflies to fly up, down, sideways or backwards, and hover for up to a minute. Some dragonflies can reach speeds in excess of 30mph. The adults of the order *Odonata* locate their flying meal by use of their outsized eyes, which can see in almost all directions at once, with each eye containing 30,000 facets. They see in colours we cannot even imagine. According to *New Scientist*:

> We humans have what's known as tri-chromatic vision, which means we see colors as a combination of red, blue and green. This is thanks to three different types of light-sensitive proteins in our eyes, called opsins. We are not alone: di-, tri- and tetra-chromatic vision is de rigueur in the animal world, from mammals to birds and insects. Enter the dragonfly. A study of 12 dragonfly species has found that each one

has no fewer than 11, and some a whopping 30, different visual opsins.

Aerial ambush of the prey is aided by the dragonfly's nervous system, which contains a circuit of sixteen neurons that allow the brain to direct the flight motor centre in the thorax. As a result, a dragonfly can lock on to a moving target, and calculate a trajectory to intercept its path. In one Harvard University study the dragonflies caught 90 to 95 per cent of the prey released into their enclosure. Prey is caught with the feet, the wings torn off with the dragonfly's sharp jaws, the meal scoffed down – all done in flight, without needing to land. *Odonata*, meaning 'toothed ones'.

The similarly carnivorous larvae (nymphs) are aquatic, sometimes for years, scuttling round their watery home by means of water squirted from the anus. As a rule dragonflies lay their eggs in water, and when the larvae hatch, they live underwater for up to two years or even more. The nymphs will moult up to seventeen times as they grow; when they catch their prey, which can be tadpole-sized, the nymphs suck the juices out to leave a hollow shell or skin.

There are about thirty-six species of the sub-order *Anisoptera* in Britain.

*

The wild celery around the pond is now two feet tall; *Apium graveolens* was used to crown the victors of the Greek Nemean games, held to honour Zeus. The pragmatic and epicurean Romans, meanwhile, exploited the herb's culinary properties; the leaves, mixed with dates and pine kernels, made a standard stuffing for suckling pig during the Empire.

Wild celery is not the easiest of plants to identify. Stems are solid, the leaves are fan-shaped and loosely toothed, and the tiny white flowers are borne in a compound umbel, similar to cow parsley. The nose detects the plant before the eye; the aroma of celery is distinct, and fills the evening.

The Farm Pond, Herefordshire
22 MAY: Aubade. The love of songbirds is very British, very old. Way back in the ninth century the Anglo-Saxons preferred to name members of the avian sub-order *Passeri* – which, strictly biologically, is what a songbird is – not by their appearance, the preference of modern science, but by their song. Finches were *finc* in Anglo-Saxon after the typical *pink pink* sound of a chaffinch. An Old English calendar for 1061 has 364 days devoted wholly to the doings of saints. The remaining day? 11 February, for which the entry reads, 'At this time, the birds begin to sing.'

Our ancestors lived by birdsong; it was their clock, as well as their calendar. They were up with the lark; they went to bed lullabied by nightingales.

We have been blessed by songbirds in these isles. Blackbird, blackcap . . . Who, having seen a goldfinch perched chanting on a spirey thistle, has not been charmed? The bird is gold-chimed of tongue, as well as gold-barred of feather.

Songbirds have been there for us in our darkest days, as individuals and as a nation. Thomas Hardy, in the 'dregs of winter' and the fag-end of the nineteenth century, was inspired by a 'darkling thrush', whose ecstatic fluting notes suggested the bird knew some blessed hope, 'And I was unaware'. When the bloody Somme battle opened in 1916 a skylark trilled sublimely over British heads, moving Sergeant Leslie Coulson to verse:

Over the troubleless dead he carols his fill,
And I thank the gods the birds are beautiful still.

He spoke for thousands in khaki.

Some spoilsport scientist will pipe that birds really only sing for sex and land. Maybe it is the fancy of frail humans to believe the birds sing for us. But listen now to the willow warbler in the alder, which surely sings for its own pleasure, as well as the birdy bare necessities.

Of course, not all birds are songsters. The heron will never be asked to join the choir.

This morning Big Bird is standing in the water; motionless, as if time were of no consequence, and nothing could break its vengeful concentration. I get to within five yards before the spell is broken, and he cranks into the air, on wings raised by pulleys.

He has been eating frogs. They are camouflaged, and are as green as the sedge in which they lurk, flicking their tongue at passing insects; frogs are able to change colour (by dispersing specialized pigment-containing cells called chromatrophores in the skin's lower layer) to suit surroundings, like chameleons, but they are mortally slow in movement.

The water roils with carp; on occasion they surface their chick-mouths open to the air to suck morsels. Pink. Lubriscous. Obscene.

In the roots of an alder a mallard has made her nest, and there she sits, as animate as a carving.

The Mill Pond, Argenton, Western France
24 MAY: 26°C in the open sun; a darter perches on a

grass stem, not along it, but head on, at right angles. Perpendicular. The sheer horizontal length of dragonflies defies physics, and aches for suspension or cantilevers.

I'm learning the difference between damselflies, dragonflies, hawkers, darters and demoiselles.

The darter, also known as chaser, and skimmer; the powderiness of its blue abdomen is 'pruinosity'; it perches on a plant to command its territory.

A dragonfly pair mating; a crooked heart attached to a ryegrass stem. I disturb them, and they fly off in the same improbable copulatory design.

Prettiest of all is the banded demoiselle, a Twenties flapper.

Pond skaters progress across the water in an insane calligraphy; the water level is sinking, so the area around the flue is a mud-bank walled by iris. The lowering of the water brings revelation: a boulder shows a shoulder, the emergency exit of the water vole is exposed. But the pond plants are at their most lush; water buttercup, cow parsley, nettles.

Dragonflies like the sun. A sunny day is a dragonfly day.

The methane bubbles from the bottom of the pond almost frantic now; in the late evening, through the wall of mosquitoes and midges, with a taper I try to set them on fire and once, childishly, manage it, for a Christmas-pudding puff of flame.

I pay a price, the traditional one at a pond. I go in over the top of the wellington, for a 'booty'.

> As when a wandring Fire,
> Compact of unctuous vapor, which the Night
> Condenses, and the cold invirons round,
> Kindl'd through agitation to a Flame,
> Which oft, they say, some evil Spirit attends
> Hovering and blazing with delusive Light,
> Misleads th' amaz'd Night-wanderer from his way
> To Boggs and Mires, and oft through Pond or Poole,
> There swallow'd up and lost, from succour farr.
> John Milton, *Paradise Lost*, Book IX

Sometimes the methane bubbles from still water catch fire spontaneously; historically the tiny flames attracted the local names 'will-o'-the-wisp', 'Jack o'lantern', 'bog-sprite', 'water-sherrie'.

Superstitions surrounding will-o'-the-wisps abounded. In Britain and Ireland, will-o'-the-wisps were ghosts of the purgatorial dead who used the flickering lights to lure travellers into treacherous marshes, thus also earning the name 'foolish fire' (*ignis fatuus*). An oral amulet was to say aloud:

> *Jack o' the Lantern, Joan the wad,*
> *Who tickled the maid and made her mad,*
> *Light me home, the weather's bad.*

In old Britain, *ignis fatuus* was also 'elf-fire' in the belief it was the work of little folk, and 'fire-drake' after the dragon's breath.

In those watery times, will-o'-the-wisps would have been more plentiful in the countryside than now, as we may deduce from Hentnzer's *Travels in England* (1598), where he relates how, returning from Canterbury to Dover, 'there were a great many Jacka-lanthorns, so that we were quite seized with horror and amazement'.

The 'lantern' lights are usually blue, but sometimes smoky white.

As Alessandro Volta, who discovered methane in 1776, proposed, will-o'-the-wisps were caused by burning methane. He thought lightning the spark; more recently it has been discovered that decomposition of organic matter in persistently wet areas produces, among other gases, diphosphane (P_2H_4), which ignites spontaneously when it meets the air. Any methane near by will be fired.

*

The Mill Pond, Argenton, Western France
25 MAY: The evening going down, the water soft, but
the ticking of the strutty little wren imitates clock
hands, a bedtime parent.

Under the sound of woodpigeons, I thought I was
in England again. The soft *coo* of woodpigeons is ever
reassuring; like the ho-hums of aged aunts talking to
themselves. Ripples of reflection climb the trees; the
same look as pulsating Christmas lights. A single cloud
is enough to end their life.

The kingfisher does not fly off from the newly
appeared boulder in the mill pond: it launches, an
horizontal projectile. Around the stone pier of the
bridge, lilies gather in the manner of open-handed
supplicants at the court of the king.

The grey wagtail is hither, thither, mind uncer-
tain, with excess of energy inappropriate for such a
languorous time, with the low *zizzing* of flies.

At the very edge of the pond down between two
stones, a frog, with a pulsating throat at the pond edge,
catches flies with its flicky tongue. Two others hop
around for no reason I can see. Courtship? It's always
the sex thing with Darwinites. Maybe the frogs are
merely having fun.

I sit here and admire the dragonflies; but insects
can be a pest. In medieval times, insects were death

and destruction. It is a modern, Western privilege just to watch them, and not see them as our murderers.

26 MAY: Leprotic yellow leaves, falling unseasonally. The algae growing exponentially.

A cuckoo, off-script; *wah-hoo-hoo, wah-hoo-hoo,* instead of the normal *cuckoo, cuckoo.*

At the pond, life never ceases. A male mallard by the alder watches me watching him. I see the curved elegance of its neck, potter-pulled.

Water crowfoot flowers, white and brilliant, an oxymoronic sprinkling of snow.

Bright Clouds

Bright clouds of may
Shade half the pond.
Beyond,
All but one bay
Of emerald
Tall reeds
Like criss-cross bayonets
Where a bird once called,
Lies bright as the sun.

No one heeds.
The light wind frets
And drifts the scum
Of may-blossom.
Till the moorhen calls
Again
Naught's to be done
By birds or men.
Still the may falls.
Edward Thomas

THE POND IN SUMMER

What wonders strike my idle gaze,
As near the pond I stand!
What life its stagnant depth displays
As varied as the land:
All forms and sizes swimming there,
Some sheath'd in silvery den
Oft siling up as if for air
Then nimbling down again.
From 'Wanderings in June', John Clare

Wandsworth, London

1 JUNE: My annual Edward Thomas pilgrimage, though this time around his childhood London, rather than the dawn land of poesy at Dymock, or the cemetery of Arras.

These days I visit the old country writers like the stations of the cross; George Orwell, Adrian Bell, Henry Williamson, John Clare, Richard Jefferies. But it's not nostalgia; it's for their records of a bird-musicked, butterfly-adorned, flowered England. Back to the future, I say.

The sun is shining on this Saturday, as I begin walking from 61 Shelgate Road, a three-storey Victorian semi, to Wandsworth Common. When Thomas was a child (he was born in 1878), the Common was more or less a wilderness, and it's on the Common that he learned his nature. Of course, in Edwardian England everyone was closer to nature. Cities were less kempt, less concreted, smaller than now. Thus, the countryside was in closer reach. When young Edward Thomas went exploring on Wandsworth Common in London it was a wilderness. The spillage of grain from horses' nosebags and its survival in their dung on every street of every conurbation was food for flocks of sparrows and finches. So abundant was the house sparrow in London's East End that it became the district's unofficial emblem. (Today, the sparrows have all but gone from London, having declined by the order of 99 per cent.) An Edwardian childhood was conducted outdoors as much as it was within the confines of the house. This was the era when every doctor and moral reformer prescribed lashings of fresh air. Cecil Bullivant's ubiquitous vade mecum for boyhood, *Every Boy's Book of Hobbies*, suggested among its pastimes 'Birds' Egg Collecting', 'The Collecting of Butterflies and Moths', 'The Making of a Botanical Collection', 'Out and About with a Geological Hammer', 'How to Make an Aviary', 'Bee-Keeping', 'Pigeons'. Edward Thomas's boyhood hobbies included keeping pigeons, and at school he wrote in his algebra book, 'I love birds more than books.' The boy next to him laughed, but only at his execrable Latin.

This happens to be the bit of London I know best; Penny's sister Kathryn lives directly overlooking the Common; Thomas would have known the house. There is enough space, few enough people, in this attractive South London suburb, with its wedding-white, stucco-edged windows, bijou shops and street trees to walk at one's own pace, rather than the speed of the crowd.

Along Kelmscott Road, over Northcote Road, a blackbird singing, someone practising violin, a nanny pushing a pram. Webb's Road, Wakehurst Road. Straight streets, thoroughly middle middle class, order-order everywhere, the four-bedroomed, bay-windowed houses in pairs. Across Bolingbroke Grove on to the Common, tree-lined, jogged, people lying in the sun reading, children playing, and the defining sound of the suburb on a weekend afternoon: 'Daddy!', exclaimed happily.

It was the ponds, really, on Wandsworth Common that magnetized the young Thomas. In his memoir of childhood, he wrote about Wandsworth Common:

Several ponds of irregular shape and size varying with the rainfall had been hollowed out, perhaps by old gravel diggings. It was marshy in other places . . . For any kind of hiding and hunting game the thickets were excellent. We played the other games in the open spaces. The ponds were for paddling in. One of them, a shallow irregular one, weedy and rushy-margined, lying then in some broken ground between the Three Island and the railway was full of efts and frogs. Bigger boys would torture the frogs by cutting, skinning or crushing them alive. The sharp penknives sank through the skin and the soft bone into the wood of the seat which was the operating table . . .

I fished for sticklebacks and gudgeon in the long pond on the far side of the railway, which owed its name of 'Backruffs' or 'Pack of Roughs', so I always thought, to the poor ill-dressed boys who used to swarm to it from Battersea on Saturdays and bank holidays. I fished with a worm either tied on the cotton line or impaled on a bent pin, and put my sticklebacks or my rare lovely spotted gudgeon in a glass jam jar . . . The gudgeon was so attractive, partly for its looks, perhaps chiefly for its comparative size, that many times I willingly paid a halfpenny for one and let it be believed that I had caught it. Even when dead it was hard to part with, so smooth and pure was it. I liked even its smell, yet never dreamed of eating it. There were carp, too, in this pond and in the roundish Box Pond that lay halfway between the top of our street and the railway. By the longer pond I once saw a carp many times as big as a gudgeon in the possession of a rough . . . Once I tried to catch these myself all alone. My expectations were huge: that I failed completely only increased my respect for the sacred pond . . . The gravelly shore of the Long Pond on Wandsworth Common was confused in my mind with the sea sand where Robinson Crusoe saw the cannibals' footprints.

The ponds – and I confess I go to Wandsworth Common in the truculent spirit of Bowling, as much as in Thomasian reverence – are still there. Box Pond, beside Bolingbroke Grove, contains two stone-lined islands, a notice about the hazards of rats, and a small pestilence of Canada geese hunkered under the trees.

Bowling would disapprove, and Thomas grimace.

On across the playing fields, to the main ponds, beyond the Gatwick Express rail cutting, with its buildings graffitied in arcane tags. 'JAFFO, USEP, EBOOT.'

There is nothing profane about the ponds beyond the railway; they remain sacred. The notice here is to inform the lucky who have walked along the board-walks through the trees that the area is an amphibian habitat. I have only entered in ten yards, when I see a father (sixties) and son (twenties) religiously taking photographs of a preening heron perched in a shading willow. Moorhens sail under the walkway bridge; there is a small raft of tufted duck out on the water of the largest pond. An elderly man, leaning forward on a walking stick, is talking extra loudly into the ear of a woman, her eyes screwed in concentration; 'The eggs have to be laid in water!'

I am happy, because Thomas died at Arras in 1917 for Adlestrop, and scenes like these. We are only required to keep them.

Because he was English, a naturalist, Thomas was besotted with ponds (like Clare and Orwell). His sub-urbia novel, *The Happy-Go-Lucky Morgans*, is proof positive. Published in 1913, Thomas himself described it as 'a loose affair held together if at all by an oldish suburban home, half memory, half fancy', about a Welsh family living in a South London house called Abercor-ran. The house boasted a three-acre garden called the Wilderness, where: 'Under the trees lay a pond con-taining golden water lilies and carp.' This is the pond, of this book's introduction, the one that needed 'nothing else except boys like us to make the best of it'.

By the time the novel begins, Abercorran House is gone but the street where it stood ('straight, flat, symmetrically lined on both sides by four-bedroomed houses in pairs') has inherited the name, and the 'three-acre field' that was its garden has likewise passed on its old nickname, the Wilderness, to Wilderness Street.

Ponds? No. 23 Wilderness Street 'probably has the honour and misfortune to stand in the pond's place'. The narrator goes back:

> The lilies and the carp are no longer in the pond, and there is no pond. I can understand people cutting down trees – it is a trade and brings profit – but not draining a pond in such a garden as the Wilderness and taking all its carp home to fry in the same fat as bloaters, all for the sake of building a house that might just as well have been anywhere else or nowhere at all.

You get the motif? Thomas, like Bowling-Orwell, goes back, and the ponds are gone.

Studies of Edward Thomas can't quite agree where exactly Abercorran House is, though it's clearly in the Clapham/Wandsworth/Balham triangle.

As I walk along Bolingbroke Grove to my sister-in-law's I see the truth of all the No. 23s, built on filled-in ponds.

Every house we build is a mausoleum for the birds and animals that lived there before.

The Mill Pond, Argenton, Western France
2 JUNE: A hoopoe, the last of the summer migrants, electronic beeping; the oriole in full lush voice; and the nightingales in the thicket along the river bank singing their tragically tender song now that 'the Queen-Moon is on her throne / Cluster'd around by all her starry Fays'.

Dragonflies: ever creatures of the sun; the sun brings them into animation in minutes; the shade quietens them. The aerial *passegiata* of the dragonflies here is one of the most fabulous things I ever saw; they proceed past my deckchair in technicolour: blue, yellow and green.

On a stick in the water, four mating dragonfly pairs perched, the male rearing from the female's back like a horse's head, or some tiny blue priapic sea-monster. But made from brittle, gorgeous Venetian glass.

Frogs squat on stones, like turd splats. Cabbage white butterflies float high in the poplars: anti-gravity white leaves.

Starling chicks in the willow, which writhes snakey to the sky; their combined voices a rattle hiss. Then one voice emerges above the others, head out of hole, it gets the grub from the parent. In bird life, those who squeak loudest get the most attention.

4 JUNE: Heavy rain, it fills and it fouls the mill pond, and a bird's eye is no advantage in muddied waters. There are six mallard ducklings on the water, line astern of their mother. A brown rat dives off the bank, swims underwater, and comes up snout-snapping. It misses – just – the rearmost duckling.

The vulnerability of floating things: subject to attack from above, below, across.

On the car radio: news report snatches from the Manchester One Love Concert, organized in commemoration of the terror attack: Robbie Williams singing 'Angels', Liam belting out 'Live Forever'.

The vulnerability of floating things.

*

Still Water

Garway Hill, Herefordshire

7 June: Climbing up Garway Hill, with its 360-degree panorama of Herefordshire and the Welsh border, the wild white ponies drinking at the pond next to the remnant ditches and mounds of the Iron Age farm. The Black Pool, which never runs dry, is a breeding area for great crested newts.

But I have climbed to look at the pond in search of archaeological confirmation. Looking at the scooped-out hollow, twenty metres in length, and intended by humans as water source for livestock, it is obvious that the Black Pool is a classic dew pond, despite being on a clay hill on the Welsh border, rather than the limestone Sussex of Kipling:

> *We have no waters to delight*
> *Our broad and brookless vales –*
> *Only the dew-pond on the height*
> *Unfed, that never fails . . .*

There was much loud literature in the early twentieth century about dew ponds, and why they never ran dry, even in the hottest summer. These small ponds can be found scattered across England, though they are concentrated in the sheepy chalk downs of Sussex and Hampshire. In the 1940s there was one dew pond per square kilometre in the Brighton area. The mystery – and the cause of the arguments and the letters to *The Times* – was how the saucer-shaped ponds filled with water. Was it dew or mist condensing overnight on the round-backed downs, as folklore maintained? Or even the hand of God, with 'dew' being a corruption of 'Dieu'?

It was the crop-circle conspiracy theory of the Edwardian era. Disappointingly for the romantics, Edward Martin, in a 1914 research paper entitled 'Dew Ponds: History, Observation and Experiment', maintained:

Dew in the strict meaning of the word can never feed a pond. It is formed from the moisture in the air being in contact with the cooled earth when this has radiated its heat after nightfall. Formation of dew on grass is, of course, a very common phenomenon. But in three months' observations on a pond there were but five occasions when the water was found to be below dew-point.

Martin was correct; condensing dew does not fill the dew pond, rainwater does. Take a look at Eric Ravilious' classic painting *Wannock Dew Pond* from 1923 (online at www.britishmuseum.org) and you will see that the pond is located towards the base of the hill, to gather the rainwater flowing down. The contribution of gravity-flowing condensed dew is unregistrably small.

Black Pool on Garway is so located, and probably dates from the same period as the homestead. The Devonian red clay on Garway is thick enough to be near-impermeable; on other geology, man-made 'dew ponds' required lining to make them watertight. In Sussex, chalk was used; this reduced to fine white dust by the driving of a broad-wheeled cart and oxen around and around the pond. According to a Sussex farmer in the 1850s:

> Water was then thrown over the latter [the chalk] as work progressed, and after nearly a day of this process, the resultant mass of puddled chalk, which had been reduced to the consistency of thick cream, was smoothed out with the back of a shovel from the centre, the surface being left at last as smooth and even as a sheet of glass. A few days later, in the absence of frost or heavy rain, the chalk had become as hard as cement, and would stand for years without letting water through.

Straw was sometimes added. Where clay was the liner, it would be 'limed' to deter the ingress of worms.

The Reverend Edgar Glanfield, Vicar of Imber, a keen student of rural life, noted the last days of the dew-pond makers, writing in the *Wiltshire Gazette*, 29 December 1922:

Up to ten years ago the dew pond makers started upon their work about the 12th of September, and they toured the country for a period of six or seven months, making in sequence from six to fifteen ponds, according to size and conveniences, in a season of winter and spring ... They travelled throughout Wiltshire and Hampshire, and occasionally into Somersetshire and Berkshire, and even into Kent.

The dew-pond maker, with three assistants, would require about four weeks to make a pond twenty-two yards – or one chain – square. He would charge about £40 for the work.

Dew ponds gradually became redundant, replaced by galvanized troughs connected to piped water.

Dew ponds possess a magic, even if the means by which they are filled is prosaic. On Garway tonight, the darkness falls only to reveal the intense inner glow of land in the hay-cut season. Windows glitter on far hills, and below all is still in the valley.

The wind, gentle and slow, slips down the hill, to help the kestrel in his work; he fans the air above the dew pond. He catches the breeze of the slope; the pond catches the rain. The ancient pond makers and the old kestrel both knew their elements.

The rising moon deepens in the sky, and a white mountain ewe comes through the sedge to sup the unmoving water. Up here above the tree-line is where farming first began in these isles. From far off, the lonely bark of a farm dog. Closer, the cawing of a ragged rook.

The breeze plays tricks, and babbles of ancient voices play in the air.

Moonlight runs down the shorn grassland, reaches the pond, and fills it with light.

The Farm Pond, Herefordshire
12 JUNE: Henry David Thoreau was of course the paladin of the pond, and one only has to commence a conversation about still water when someone asks, 'Have you read *Walden?*'

Today I finished with Thoreau's *Walden*, as in closed the book but also as in the termination of seeking inspiration in its pages. I fuelled my adolescent anarchism on *Walden*'s libertarianism (a decent PhD could be gained on *Walden* as the first teen-angst text); when living for a year on wild food I was succoured by his self-sufficiency lessons; but today I cannot get past the man's egotism. I will never again wade through *Walden*.

Henry David Thoreau was born David Henry Thoreau in 1817, the third of four children of a pencil manufacturer in Concord, Massachusetts. After graduating from Harvard, he worked as a schoolteacher. He gave up teaching after a couple of years, but only in the classroom; his books are essentially exercises in pedagogy.

Thoreau went to live on the pondside at Walden, Massachusetts on 4 July 1845. He spent two years at Walden but nearly ten years writing *Walden*, which was published in 1854 to middling acclaim and sales. Only after Thoreau's death in 1862, and thanks to energetic championing by his friend Ralph Waldo Emerson (who actually lent the land at Walden to Thoreau), did the book become the bible of anti-authoritarianism, survivalism and conservationism.

Thoreau went to Walden Pond, in his words, 'to learn what are the gross necessaries of life', or the fundamental basics of being human, unencumbered by accoutrements and fleshy temptations. One might cavil that his attempt to live off-grid was rank hypocrisy; he walked home to get cookies from Mom in Concord at least once a week, and Walden Pond, far from being isolated, had the Boston railroad running along one bank, and in summer heaved with picnickers. In winter it was frequented by ice-skaters and ice-cutters.

For others he advised an asceticism so stringent as to make a Benedictine monk look like a disciple of Bacchus; for himself, he prescribed dinner parties in his cabin. Thoreau's self-imposed exile existed only in his head.

Walden is a state of mind.

And that ego of his. In the catalogue of landscape features nothing is more modest than a pond, but the quality did not apparently touch Henry David Thoreau. A single sentence from *Walden* will suffice. 'Sometimes, when I compare myself with other men, it seems as if I were more favored by the gods than they, beyond any deserts that I am conscious of; as if I had a warrant and surety at their hands which my fellows have not, and were especially guided and guarded.'

I suppose he did have much to brag about, and I do not want to leave *Walden* on a low. In musing on the land and our relationship with it, Thoreau was properly a seer. 'We can never have enough of nature,' he wrote. 'We need to witness our own limits transgressed, and some life pasturing freely where we never wander.' However fake was his own hermitage he understood why the countryside and the wild places matter, and why the blood is livened when one is alone with nature.

Thoreau's descriptions of nature in *Walden* have an energy, an original way of looking, unlike his passages of philosophy, which never lift above the 'ye shall not' jeremiads of the minor prophets of the Old Testament. He is beautiful and accurate on the breezes, bugs, birds of Walden, and the ice of winter:

I sometimes used to cast on stones to try the strength of the ice, and those which broke through carried in air with them, which formed very large and conspicuous white bubbles beneath. One day when I came to the same place forty-eight hours afterward, I found that those large bubbles were still perfect, though an inch more of ice had formed, as I could see distinctly by the seam in the edge of a cake. But as the last two days had been very warm, like an Indian summer, the ice was not now transparent, showing the dark green color of the water, and the bottom, but opaque and whitish or gray, and though twice as thick was hardly stronger than before, for the air bubbles had greatly expanded under this heat and run together, and lost their regularity; they were no longer one directly over another, but often like silvery coins poured from a bag, one overlapping another, or in thin flakes, as if occupying slight cleavages. The beauty of the ice was gone . . .

And I confess I have, over the years, rather identified with Thoreau's inability as a fisherman, and his conversion from wildfowling:

> As for fowling, during the last years that I carried a gun my excuse was that I was studying ornithology, and sought only new or rare birds. But I confess that I am now inclined to think that there is a finer way of studying ornithology than this. It requires so much closer attention to the habits of the birds, that, if for that reason only, I have been willing to omit the gun.

13 JUNE: Usually I write my books on scraps of paper; I'm writing this next to the farm pond, laptop on seated lap, the writer's equivalent of the artist at his/her easel painting the landscape.

The water is beery-brown; the water lilies are the definition of tranquillity. Why? We fear sinking, while water lilies float effortlessly on still water. Why do you think Monet painted them over and over again?

*

In contrast to the constant nervy movement of the flies around the ponds, the heron is a pole of stillness; the dragonfly is both a mover and a statue. For a second, the dragonfly is poised in the air, an insectoid hummingbird, moving/unmoving, then darts away, to resume its strained aerial repose a few yards further along the sedge.

Under the water, peering close, at the edge of a stone: the tiniest wag of tail. Bend to lift a stone. A fog of sediment arises. Make a grab, to triumphantly hold aloft a common newt.

Big hatch of mayflies. Green drakes and pond olives (*Cloeon dipterum*) mostly, swarming under the pond-side trees from mid-morning.

The dance of the mayflies, a summer dream. How strange their life is.

For up to three years these fairy-like creatures have been living a worm-type existence in the mud of the pond, feeding on algae and plants; millions of other mayfly larvae are down there now, heaving, waiting for a glorious summer day next year. Two years for the mating dance in the sunshine, which may last only hours.

When the sun sinks they will die. Many species of mayfly do not feed as adults; their sole purpose is to reproduce, dying once they have mated. They don't even have functional mouth parts.

Later: hundreds of mayflies fall exhausted on to the water. I lift one out, blow on it gently; they are delicate creatures, with broad, clear wings that have a lace-like appearance, and three fine tail bristles. The mayfly re-flies but only back to the pond, to collapse again, and die.

The name 'mayfly' is misleading as many mayflies can be seen all year round, although one species of the fifty-two present in Britain, *Ephemera danica*, does emerge in sync with the blooming of hawthorn, or 'mayflower'.

Mayflies are unique as insects in having two winged adult stages. After emerging from the water as nymphs they fly to the bank where they shelter on the underside of leaves or in the grass. They then moult again, leaving behind their drab 'dun' skin to reveal their shiny 'spinner' skin.

Mayflies were one of the first winged insects; their fossils date back over 300 million years – long before the dinosaur appeared.

The next morning it rains, and the raindrops pit the water in elven-hammer blows, sinking the mayflies and the flotsam, to clean the surface of the pond.

*

Big hatch of ducklings. The mallard under the alder has six of them, which take to the water . . . well, like ducks.

An email from Australia from my childhood friend, Tim, who I'd asked about pond fishing in Hereford-shire in the 1970s/80s.

Growing up on the banks of the Wye and a little fur-ther the Lugg, pond fishing never seemed like an option until the day 'Specky Stevens' said, 'I know where we can catch roach!' Specky was a gangly kid one year older than myself and was not renowned for his angling skills. However, his father was a different breed of creature. I can still see his garage, neat and tidy with a forest of rods all along one wall. Now if Specky's dad said he knew about roach then this could be worth it. This gent had guided me one weekend along the banks of the Pandy brook where I caught my first grayling.

Why the attraction of roach when you had the Wye to fish? Mr bloody Crabtree, that's why! In his book was a full page entitled 'The Roach, the Most Popular of the Fish'. With a plate showing several of the silver red-finned beauties swirling through a weed bed. They may have been the most popular fish but in the 1960s there were none or very few in the Wye, which was dominated by dace and chub. Intrigue got the better of me and so on a fine summer's evening four of us set off on our bikes, rods tied to the crossbars and tackle bags hanging off our shoulders as we followed Specky Stevens into the unknown lanes of Herefordshire.

After what felt like hours, much moaning and cussing we arrived on the outskirts of Withington. Down another lane and behind a hedge we glimpsed the glistening of water. Once parked up and through the hedge our hearts sank. What seemed no more than a puddle festooned with broken branches was Specky's roach mecca! 'How the hell are we going to fish this,' we exclaim to Specky, who at this point could feel that he had led us on a merry goose chase. He just shrugged and suggested we give it a go. After a while the boyhood disappointment died down and the spirit of adventure made us take in our surrounds. First one and then another fish was seen to rise. There is life in this pond! Encouraged, we tackled up and soon had learnt this was not fishing the Wye with wide open banks. Techniques had to be adapted and after a few cases of float versus tree encounters we mastered the art of fishing in a very complex environment. Floats started to dip and hearts started to pound but the roach was obviously a much more subtle feeder than the chub. But before long the pond gave up its silver treasure and we were now adding roach to our catch list. I say roach, well they were only two to three ounces but to us they were the new mega fish. By dusk, Specky had deserved his reprieve and four happy boys cycled home in the dying light with that certain feeling of satisfaction only young anglers know.

My other experience of ponds in Herefordshire was at Belmont opposite the school. Down in a red clay hollow surrounded by oaks lie the remains of several Friday pools. These were the fish hatcheries set up by the monks of Belmont Abbey to ensure they always had a good supply of fresh fish for their Friday meals. Again, Herefordshire was not known for its carp (with the exception of Redmire, the secret location of Britain's record carp for many years), but tucked away, next to a main road, there was a source of crucian and common carp. My good friend, angling nemesis, best man and godfather to my eldest, Mike Prosser, had heard about this place and was keen to show it to me. Now a little older and a little wiser we approached the ponds with caution, more because the wet red clay was extremely treacherous under foot. Swims were selected, tackle set up and arranged and bets laid. The latter normally involved beer after the fishing and even thirty years later I very rarely do not get to buy the first round! I am sorry to say that even with maturity the feeling of watching that float dip whilst trying a new water and different species never wains. It certainly did not this day as we chuckled placing fish after fish into the keep net, with the exception of the larger fish that were weighed and released as soon as possible.

Over the years I have now fished many ponds around the beautiful county of Herefordshire. But nothing beats the days when you get to fish that tucked-away water for the first time.

The Farm Pond, Herefordshire
14 June: Violence in the air, the clouds thundery; a kestrel takes a moorhen chick off the face of the pond.

Skitty's chick was in the duckweed, which stretches out in a green raft from the south bank, and I wonder if the kestrel – not usually a water bird – thought the duckweed lawn. Skitty has one chick left. Four have been lost to time without record, which is the lot of almost all of us, birds, beasts, humans.

Of the six mallard duckings, only three remain; every day their family caravanserai becomes shorter and shorter.

Hydroponics – the nurseryman's art of growing plants in nutrient solutions without the need for soil to root in – was discovered aeons ago by duckweed, which floats rootless in ponds. The most abundant of the family in Britain is lesser duckweed; greater duckweed (*Spirodela polyrhiza*) has significantly bigger leaves, as its name suggests, and these are usually purple underneath; ivy-leaved duckweed is common in the south; rootless duckweed, with its leaves 1mm across, is Europe's smallest flowering plant (*Wolffia arrhiza*) and frequently mistaken for a scum of green algae.

I try a swim. My method of entry into the water is that described of the Englishman in Fitzgerald's *Tender is the Night*: 'With much preliminary application to his person of chilly water, and much grunting and loud breathing.'

Pond-dipping, really trawling the bottom mud, each scoop-up bristling with larvae and bugs, monstrous in their ceaseless purpose; tubular larvae of crane fly; the

Halloween-headed nymphs of gnats; the millipedey-larvae of whirligig beetles; the shrimp-like young of mayflies. In one square foot of pond, I find over sixty insect larvae.

In this bottom zone of the pond oxygen is endangered, but the 'bloodworm' larva of the midge possesses the blood pigment haemoglobin, which, on account of its affinity for O_2, enables it to breathe in the smothering morass.

More than a hundred pond skaters skittling about. Of course, in summer the pond is far from being still water. The pond vibrates with life. In the weedy edges, button-sized froglets seethe in their hundred hordes. It has taken three months for the tadpoles to complete their metamorphosis, to become clone-miniatures of their parents.

15 JUNE: Each month brings its own beauties; I never tire of watching the great galleon clouds of June. Lying on the ground, next to the pond, the picture of idle countryman, I see the hills dissolve in the distance. Heat and light on the reflecting water glow on my face. The pond skaters set about their business, walking on water; float-flick, float-pause, float-flash; always

smoothly gliding, always their splayed feet leave club-
foot shadows on the pond bottom, and only the vaguest
imprint on the surface of the water. They can see their
surface prey, but also sense vibrations. They have
already mated; within weeks the new generation will
be hunting the pond.

Pond skaters are creatures of still water. When the
surface is ruffled they stay in their harbours amongst
the vegetation and stones. It is a pleasing pastime
to have lunch – if only this sandwich – on the pondside, studying the delicate lines, circles and patterns
inscribed on the polished surface of the water by the
skaters.

The pond: the moorhen's bath, the heron's paddling pool, the pond skater's rink.

A carp rises, jarring the surface, but the disturbance
is smoothed away gently by the boundlessly forgiving
water.

The clouds are exactly reflected in the water, as is
the breast of the blackbird, as she crosses and re-crosses
the pond to her chicks in the alder.

Blackbird is the surprising unsurprising pond
bird, probing the mud as equally as she probes the
lawn.

As the sun's rays soften, I tire of the languidness of
the water and am about to leave to go back to the

tractor when a hawk flickers over the pond, a winged flame, and vitality returns.

Tonight after supper I blow up the Sevylor canoe, carry it on my shoulder (it has no real weight; it's inflatable), put it on the water, clamber in, and row to the centre of the pond, which takes about five seconds. It is one of those magical June evenings one gets after a hot day, absolutely still with no hint of breeze, and all the trees laden down with their hardened foliage.

The canoe is as stationary as if moored; I ship the oar, lie back, a leg up on each side of the canoe. There is in the air a quite fantastic potpourri of grass, elder-flowers, dock leaves, reeds, lilies, and the pond's own special flavour, a wild lustral tang that is somehow intensely exciting to me. Mayflies dance in fountains. At this time of year the birdsong is fading but the blackbird performs evensong in my personal church. Some early hornets are droning about. The mallard and her young swim out from the reeds, the ducklings line astern, cross the pond – I could touch them if I chose – and then scamper up the bank.

I cannot tell you how at ease I am, alone in my own little boat, with just the rubber skin beneath me so that I feel every undulation, ripple. Like a duck.

A willow warbler begins to sing in the alder above the lovely jungle of reeds, water crowfoot and watermint. And the spotted flycatcher dives off her twig, catches some nameless winged bug over the water. She suits the habitat exactly. She is a fisher of insects.

The Mill Pond, Argenton, Western France
17 June: Eradicated by flood, submerged in cannibal fury, the pond becomes river.

18 June: The first to re-establish on the pond are the skaters, tens of them in hunting packs, all facing the same way into the breeze, and all in the patches of sunlight; constantly shifting their position so they remain on station.

A hatch of pond skaters *Aquarius najas*; at 17mm, bigger than the usual *Aquarius paludum* with its yellow go-faster stripe. Watching them disturb the surface of the pond, I note that their Subbuteo-flick produces a linear series of overlapping circles.

Fe-fi-fo-fum, the tiger mosquito likes the blood of this Englishman, and I'm bitten on the arm, and contract some sort of infection that leads to it doubling in size, as if I had spotty dropsy.

This invasive species (it is originally from Asia) is a vector for serious diseases, including dengue fever and viral infections chikungunya and Zika.

In certain communes in France new predators have been introduced to try to limit insect numbers, including fish that eat tiger mosquito eggs, and bats that can eat up to two thousand mosquitoes in one night.

Dragonflies also do their bit, eating hundreds of mosquitoes per day.

The Farm Pond, Herefordshire
22 JUNE: Scented water. The clumps of watermint in full noisome glory. 'The savour or smell of watermint,' wrote herbalist Gerard in the sixteenth century, 'rejoiceth the hart of man for which cause they strowe it in chambers and places of recreation, pleasure and repose.'

Watermint has given its name to Minety in Wiltshire, and Minstead in Hampshire. The small lilac flowers of the watermint know only summer; they appear in late June and are gone by September.

The taste of watermint is harsher than that of the garden's common mint, but can be used for all the same sauces, chutneys, cold drinks and Moroccan-style teas. I pick some to add to new potatoes.

Up the long leaves of yellow flag, dragonfly larvae climb, the leaves their ladder to the sky.

My entrance across the field takes Skitty by rare surprise; she is on the open water with her one remaining baby. Skitty scrambles over the water for the briar; the chick submerges – it absolutely does not dive; it sinks like a lead brick. I spend five minutes waiting for it to come up. Again, I lose the game of spot the bird underwater.

Or, mostly underwater. Moorhens can lie nine-tenths submerged, with only their head showing. 'Dip-chick' was one old, and accurate, country name for the bird.

The heat has evaporated a full foot of water, leaving a mud band around the pond; the rim in the bath. Big Bird the grey heron came in the night, betrayed by his outsize footprints in the ooze. Even against my size ten boots, Big Bird's feet measure up impressively; they are 5.5 inches from the tip of the front claw to the tip of the back claw. The length and the span of the feet prevent the four and a half pound bird sinking in squelch.

There is another impression in the morning mud: that of a rodent's body, a vole or a mouse, something of that order. Herons like their mammalian meal motionless, so they stab it in the head with piston-regularity until life has left it. Since prey with fur is no easy swallow, the heron has a wise ancestral trick to help the animal go down, which is to dunk it in water to moisten the fur, make it slippery, slidey.

That heron you saw standing in a field was perhaps not hunting rodents and frogs, but digesting, because the heron can consume prey as big as adult sea trout. Sometimes several digesting herons will stand together, in a mutual silent, still ceremony. Herons are impressive as well as impassive hunters; the classic study of herons, *Foraging Behaviour and Food of Grey Herons* Ardea cinerea *on the Ythan Estuary* by D. C. Cook, determined that an adult heron is successful with 50 per cent of its catching attempts.

Fish compose the bulk of a heron's diet, but *Ardea cinerea* is no faddist, and will take anything that is alive and can be swallowed whole. Ducklings, wader chicks, frogs, shrews, moles all feature on the heron's menu. On this farm pond, the grey heron has taken newts and diving beetles. Prey is swallowed whole but digestion is so industrially efficient that only a grey paste is present in faeces. Indigestible elements, such as chitin, fur and feathers, are cast up in oral pellets.

The heron's domain is mud and still water. To keep clean in the mire the bird has evolved special feathers on its breast, which it crushes with its feet into granules and spreads over itself. This 'powder down' soaks up the muck and grime from its feathers, which it then scrapes off with a serrated claw.

Given its strangely human looks, it's small wonder that the heron is steeped in folklore. Once upon a time anglers believed its feet gave off a scent that magnetized fish, so carried a heron's foot to bring them luck.

The Old English name for heron was *hragra*; other names now largely fallen into obsolescence include harn, moll hern, hernser, hegrie and hernshaw. Heron comes from the French; the Gallic name is *héron cendré*. All of them, of course, are superior to the scientific name *Ardea cinerea*; ornithological scientific names are a useful universal language, Esperanto for the birds, but they lack poetry, and the ability to fire the imagination. Such as the Pembrokeshire name for *Ardea cinerea*: longie crane.

Shakespeare asked, anticipating French critical theorists of the Lacan ilk by four centuries, 'What's in a name? That which we call a rose / By any other word would smell as sweet.' It is an odd remark from a litterateur whose stock-in-trade was the descriptive exactitude of individual words. In *Hamlet* the prince raves: 'I am but mad north-north-west: when the wind is southerly I know a hawk from a handsaw.' Handsaw was a northern folk name for the heron, and Shakespeare counterposes the nobility and sharpness of the hawk with the plebeianism and roughness of a tool.

Like soil samples brought up by the geologist's drill, country names record the creativity of the British peasant down the centuries. The apostle of the folk name was the nineteenth-century rustic poet John Clare, who sent a sharp retort to his publishers when they queried the dialect term 'woodseer' for froghopper: 'Whether it be the proper name I don't know tis what we call them and that you know is sufficient for us – they lye in little white notts of spittle on the backs of leaves and flowers. How they come I don't know but they are always seen plentiful in moist weather – and are one of the shepherds weather glasses.'

Woodseer, meaning 'wood prophet', captures precisely the insect's weather-sensitivity. Woodseer, that single word, connects us both to the animal and to the society that coined it.

Clare's determined desire to accurately record his

shrinking world means he himself avoided neologisms. But the nineteenth century was the baptismal age of British natural history, with parson-naturalists and enthusiastic amateurs rushing around in a great naming of names. This was the period when many of our butterflies and moths got their standard common names. Red admiral was conjured because the patterning of its wings (patches and stripes) looks like a naval ensign. The Mother Shipton moth bears the image of the eponymous sixteenth-century hook-nosed witch on its wings.

At night: the toads are leaving the pond in their ones and twos; there is not the same mass movement away from the breeding waters as to it. The meadow grass is long, and wraps itself around the pond, the night, the toads.

Edinburgh
30 JUNE: There's a Jewish joke, where a mother says 'Help! Help! My son the doctor is drowning.'

Anyways, we're in Edinburgh to attend the graduation of our son the doctor, and are booked into the Ibis at Edinburgh Park, this being one of the few dog-friendly hotels I can find. The hotel's address is

6 Lochside, and I'm expecting an utter fiction of an address, in the way that Acacia Avenue is entirely free of acacias, and Kingsway Shopping Centre will never be troubled by the footfall of monarchs.

'Lochside' is not a total lie, more an estate agent's sleight of Word 2016. There is no loch, but there are two interconnected ponds which make a literally concrete case of the snobbery in the naming of watery bodies. JP Morgan and HSBC would be unlikely to have their glassy, sun-shiney Edinburgh offices on 'Pondside', or Ibis locate a hotel there. Estate agents, however, are far from being the only offenders in the 'euphemizing' of still waters, and literature is full of it. In Greek myth, Nemesis leads Narcissus to a body of reflective water, invariably translated as 'pool', where he suicidally falls in love with his own reflection.

The prosaic truth is that Narcissus, son of the river god Cephissus and the nymph Liriope, looked into a pond.

Whatever the semantics of Edinburgh's 'Lochside', the two artificial rectangular ponds fronting the commercial development brim euphorically with flora and fauna. There are bands of bulrushes ten metres thick; on most rushes, the familiar sausage-shaped seed heads have split apart, spilling the cotton-light seed. A few bulrush heads have been doused by rain mid-eruption, then been baked by the sun to become solidified fawn candyfloss sticks. There is water buttercup, yellow iris, rosebay willow herb, horse chestnut, yellow lilies in their full drinking-chalice magnificence, and alders with the cones at the little green grenade stage. In the water, a shiver or two of fish, roach I think.

The pittery pied wagtail on the paving is no surprise, nor the three moored mallard on the water, nor the squeaky moorhen well-blinded by reeds. That a heron works the gravel-bottomed ponds I would not have guessed; we meet by accident, as I amble along the paved path in the foreground of an empty, seagull-crowned office block. He is down in the reeds, smuggled, snuggled by blinding light on the water. Infiltrated. Front on, the grey heron is all white; I presume the creatures he stabs, if they chance to look up at his looming-over breast, see only familiar English cloud, a pictorial analgesic.

We are so very proximate I can see myself in the round eye when his head turns away; it's a social awkwardness, the one where you met someone at a supper years ago, so do you acknowledge or cross the street? The heron decides the latter, and rises up, but to be instantly mobbed by two herring gulls coming off the cliff face on Number 1 Lochside, their squawks cannoning off the corporate temple; the heron shrugs the gulls off, skylines towards the other 'loch', only to be dive-bombed by a carrion crow. The black bird is scolding, crow-pecking; my heron swerves a lateral left, as ungainly as blown rags.

I am watching a slow tragedy unfolding, a heron-crash, of my own making. Big Bird is now flapping amid the wires of the Edinburgh tramway; a wing catches a line, sufficiently for him to tilt, sufficiently for feathers to fly.

My hands to my face, peeping between fingers. My heron, my heron. Tumbling down to the rails, the heron rights, climbs the air. Now he is magnificent: in his space, the natural sky.

Some Sunday office staff lying on the clover banks in the Seurat haze applaud.

'Bulrush', *Typha latifolia*, I should add, is a misnomer inherited from Victorian times, when Sunday school books pictured 'Moses among the bulrushes'. The true bulrush is actually another species entirely, *Scirpus Lacustris*.

The bulrush in Edinburgh, and on the farm pond, is properly reedmace, and is a forager's wetland dream, proving edible succulent shoots, rhizomes (roots) and flowers. There's no time of year when reedmace does not provide something for the table. And it is abundant and usually easy to harvest.

Look for tall belt-strap stems (sometimes reaching to more than three metres in height) and a velvety brown flower spike at the edges of swamps, ponds, lakes and slow-moving rivers. To harvest the thick, rope-like roots that grow underwater, follow the stem down into the murk with your hand and excavate some of the mud away. Then grab the root and pull up hard – either a whole root or section of root should follow. Roots are at their fullest in autumn and winter. Washed, roots can be eaten raw, though roasting in the manner of yams makes them sweeter and more digestible. Hunter-gatherers through the ages have also pounded the roots to release their starch in a versatile, highly nutritious and digestible form: flour.

Reedmace shoots, which grow off the tubers, can be one of the first signs of spring, bursting up with all the vigour of bamboo as early as February. Washed and stripped of their brown outer leaves, the shoots make a fresh, crunchy wayside snack, not unlike asparagus. Take them home and stir-fry in a wok, or steam like asparagus. They taste of sweetcorn. Actually, the job is better done at home because peeling off the outer layers of leaves releases a sticky jelly that requires much washing and scraping of hands. At home, the jelly can be collected and used as a thickener like carragheen.

There is yet more of reedmace to be eaten. In summer the pollen is prolific enough to be collected and used like sweet flour.

This six-foot sentinel of the muddy pond edge was, in medieval times, harvested on the Eve of St Brigid's, 31 January. (Strictly without use of iron-cutting implements, which brought bad luck.) The leaves were woven into small crucifixes known as St Brigid's crosses, believed to prevent spirits from entering the house. In February the reedmace sheds cottony parachutes of seed, housed within the familiar padded brown seed head, into the cold wind of winter for dispersal.

The Farm Pond, Herefordshire
5 JULY: Back in the swim; the overhead curved world blue, the flat water pale brown; my arms appear in sepia photo-flashes as they come together to begin each stroke.

My long-held thought that the pond is best understood in cross sections is to be put to the test: I have done the surface, and the exact line where air meets water across the eyes; now is the time for underwater. I edge into the pool, feet probing forward, through the drowned, blackened sticks and leaves, silky seductive, the mud sensuous, like putting one's feet in satin slippers . . .

My earliest memory of swimming: a red badge for managing the length, chlorine and choking in the school swimming pool aged seven; then nothing until my early teens, and the sybaritic pleasure of the private use of Bishop of Hereford's Bluecoat School's outdoor pool, all warm concrete slabs and blue water and chlorine, after tennis; a bit of Hockney in Herefordshire. (My stepmother was a French teacher at the school; we had the key to privilege.) I am anxious in water, unlike my lamented aquatic father, player of water polo, rower, sailor, who performed the crawl as though modelling for Ancient Greek sculpture.

. . . I go under, immediately up on my feet. The blood screams. Then I strike out with broad shoulders, my ear full of my admonishing father (rightly too), four quick strokes, then face down.

It is pondy Piccadilly Circus down here. An adult newt, type indistinct; thick krill-shoals of mosquito larvae; beetles; toadlets; then, out of the murk, the dread hull of a carp lurches towards me. There are already things crawling on my body . . .

It is good and it is exhilarating, and I want to do it again.

*

STILL WATER

That adult newt is the guest who stays too long; all the other adult newts have left the pond to begin their terrestrial phase; they won't return to the pond again until next year and the breeding time.

The recent sunny days have caused the water to fall by three or more inches, and on the exposed mud-ring the paw marks of the fox are starkly obvious. A scattering of buff feathers, as though proffered in a taunt, a vulpine dropped gauntlet.

One duckling still swimming.

The New Forest
Summer holiday, the remote past: A picnic, overflown by dragonflies (pleasing) and bluebottles (nauseating); I walk to the pond, under trees, in 'jelly sandals', edge slightly in – a writhing black torpedo appears in seconds, then latches on to my flesh. A medicinal leech. I shout, kick out; the leech flies off; but there is blood, which mingles with the milky water.

I was lucky in this encounter. Medicinal leeches (*Hirudo medicinalis*) are rare in modern Britain. Save for this moment we might never have met.

*

To understand the present straitened predicament of the medicinal leech in the wild we need to go back to William Wordsworth. In 'Resolution and Independence', 1802, Wordsworth – surely the ultimate example of nominative determinism? – described an encounter he and his sister Dorothy had with a leech-gatherer beside a moorland pond:

> *He told, that to these waters he had come*
> *To gather leeches, being old and poor:*
> *Employment hazardous and wearisome!*
> *And he had many hardships to endure:*
> *From pond to pond he roamed, from moor to moor;*
> *Housing, with God's good help, by choice or chance;*
> *And in this way he gained an honest maintenance . . .*
>
> *My question eagerly did I renew,*
> *'How is it that you live, and what is it you do?'*
>
> *He with a smile did then his words repeat;*
> *And said that, gathering leeches, far and wide*
> *He travelled; stirring thus about his feet*
> *The waters of the pools where they abide.*
> *'Once I could meet with them on every side;*
> *But they have dwindled long by slow decay;*
> *Yet still I persevere, and find them where I may.'*

The words of the leech-gatherer in Wordsworth's poem suggest that the species was already in decline by the beginning of the nineteenth century, almost certainly due to over-extraction from the wild. The medicinal leech – which reaches up to 20cm in length, and is dark green with reddish longitudinal lines, despite my memory of them as *noir* – is the only British leech capable of sucking blood from humans. When Wordsworth was writing, the trade in leeches was approaching its zenith: by the Victorian era, Britain used over 42 million leeches a year for phlebotomy, or bloodletting. The European medicinal leech was preferred over its American counterpart (*Hirudo decora*) because it consumed larger amounts of blood.

There were few ailments for which leeches were not employed. The belief was that the medicinal leech had the ability to extract 'bad blood'. It was an ancient belief: medicinal leeching is depicted in Egyptian paintings *c.* 1500 BC; the Talmud, Bible and other Jewish manuscripts outline the medical practice of leeching. The Old English word for physician was *laece*, indicating that doctors and the annelids were etymologically overlapping.

Leeching was less painful than bloodletting by the lancet or the cupping glass, thus people would pay for treatment by the sanguivorous ectoparasite.

Leeches were big business. Despite the Wordsworths' meeting with the gentleman leech-gatherer, traditionally leeches in Britain were harvested by women, who would lift their skirts and wade into ponds and mudflats, allowing the leeches to attach themselves to their bare legs for long enough to take them to an apothecary or a leech dealer, who would transport the annelids around the country on the body of an old horse.

While the women themselves would be lucky to earn a penny per leech, doctors would charge six times as much for their use on patients.

By the end of the nineteenth century, leeching had fallen into disrepute as quackery, but there was no respite for the medicinal leech. The medicinal leech requires relatively high temperatures, particularly for breeding, and is typically found in shallow water with plenty of submerged and marginal vegetation, where above-average water temperatures are maintained in the spring and summer. Ponds and pits, in other words. When the pond was ploughed under, built on, silted up, the medicinal leech perished with it.

By the beginning of the twentieth century, the medicinal leech had disappeared from most of its former range in Europe and was declared extinct in the British Isles. Since 1970, however, populations of *Hirudo medicinalis* have been found scattered across the British Isles, with the gravel pits and ponds of Dungeness in Kent being particularly populous.

The twist in the tale of the medicinal leech is that, although it remains a threatened species in its natural habitat, it has made a comeback in medicine, and is being mass-reared for this purpose.

Leech therapy is now used by the NHS to restore circulation to grafted tissues and reattached appendages. Medicinal leech have over three hundred tiny teeth in three sets of jaws, which they latch on to their host. As they feed, they apply the perfect amount of suction to restore blood flow after delicate reattachment surgery. In half an hour, a leech will take about five times its own weight in blood, amounting to between 5 and 10cm^3. As many as fifty leeches may be used in succession on one patient post-operatively.

Medicinal leech saliva also contains medical compounds that have anaesthetic, vasodilator, anticoagulant and clot-dissolving properties.

Leech therapy has established itself as an alternative remedy for the treatment of vascular disorders. In 1997 an antithrombotic and anticoagulant pharmaceutical preparation was released to the Russian markets under the trade name 'Piyavit', which consisted of medicinal leech saliva extract. Leech saliva has also been found to be antimetastatic and analgesic.

The quacks who employed leeches were not so wrong after all.

A beech wood, Carmarthenshire, Wales
The remote past: We stopped by the side of the road, because the car was overheating, and I wandered off into the adjacent beech wood. I was about fifteen. Beech-skin is made for carving, and I etched my initials, and those of a girl I briefly fancied. (The beech tree is still there and likely will be for a hundred years more; we cut down trees, they cut us down to size.) Further into the wood, full of shade and silence, I came to a black pit, a giant's grave, where all the leaves of the decades had mouldered into vile, anoxic tar. Standing on the side, I poked the mire with a stick; but the stick was not long enough. Another stick was fetched. The pit was four feet deep; the smell was the musk of all the dead, ever.

I thought at the time the pit was some abandoned vehicle inspection pit (we had a similar one at home) but later came to realize it was a saw pit, where logs were sawed lengthways with a two-man saw, one man standing down in the pit. The practice appeared in the fourteenth century.

Saw pit ponds are common (when you know what to look for), and even in these mires there is life. I half expected Golum at the bottom of the Carmarthenshire pond, and in a sense I found him. A pale worm, *Haplotaxis gordioides*, wriggled out. It was more than a foot long. I know because I had my *Observer's Book of Pond Life* with me, which had a handy imprinted ruler on the back cover.

'Wood' in Medieval English had the secondary meaning of 'mad': Lysander in *A Midsummer Night's Dream* complains of being 'wood within this wood'. The white worm in the Welsh wood made for mad nightmares for years.

The Mill Pond, Argenton, Western France
8 JULY: The trees drenched in the chorusing of willow warblers; the male golden oriole still singing, now answered by the parrot-puke call of the female.

The seed heads of the yellow flag exactly like small green capiscum. I give the velvety head of the bulrush an affectionate squeeze in passing.

The beautiful epilepsy of five banded demoiselles on blue wings.

The pond's banks in the afternoon: full of flowers and murmurs.

*

The Farm Pond, Herefordshire
10 July: My friend the writer David Hill has sent me a copy of *I-Spy in Pond and Stream* to aid my endeavours. Produced by the *News Chronicle*, *I-Spy* books were all the rage in the 1950s, 1960s and 1970s. The pleasure was simple: spy something, learn something, earn some points. In the *I-Spy in Pond and Stream* there were a total of 1,500 points to be collected, and on gaining these one sent off one's little paperback book (one shilling) to Big Chief I-Spy for an Order of Merit.

I never did the pond *I-Spy* in my childhood, when David Bellamy was the Big Chief, so I am intending to make up for lost time. Big Chief advised:

> Sit quietly by the water; watch the miracle of the dragonfly emerging from the larva; the water boatman rowing on his back; the young eel in the last stage of his fantastic journey . . . With *I-Spy in Pond and Stream* in your pocket set out for your nearest water 'hunting ground'. You will find it an intriguing trail.

I do as Big Chief I-Spy, Head of the Redskins, requires (the racism already jarred in the 1970s; the hippie elder siblings of my friends were all reading Dee Brown's *Bury My Heart at Wounded Knee*).

In ten minutes by the farm pond 'I-Spy': daphnia or water fleas (15 points), and freshwater louse (25), pond skater (15), pond snail (10), curly pondweed (15), bulrush (15), lesser waterboatman (20), whirligig beetle, 'a whizzing ball-bearing' (10), frogbit, 'a miniature water lily in appearance' (10), lesser duckweed (15) and a whopping 35 points for Daubenton's bat, which comes out early to help my total.

The little book gives a sad and arresting insight into the decline of the water vole. In the 1950s, the water vole was so utterly common it merited a mere 20 points. Today it is endangered, and absent from whole tracts of the countryside,

The *I-Spy* books work well for the competitively minded. I really want to see a red-tailed maggot (25) and long water scorpion (25), and put both on my sand-bucket list.

The Farm Pond, Herefordshire
10 JULY: Across the meadow, paddling through the knee-high buttercups; the way ahead dissolved in mirage. Only in the last twenty yards do the vibrating motes cohere into a pond with trees, reeds, dragonflies.

Dreamy, Grecian pond. It is a revelation, these daily changes. On this afternoon there should be naiads and fauns reclining around my rustic pool. The water is velvet comfort; and the air, each atom of it, upliftingly light.

High summer: the weed is bursting, the flowers at their most luxuriant; the adult amphibians mostly gone, the carp retired to the deep cool under the briar. (Where I prefer them to be; fish are meant to flit and hide from humans; the carp's open muscly cruising is contrary to the laws of nature.) A dragonfly nymph climbs laboriously out of the water, up the spire of a reed. I arrive to see a dragonfly, just birthed from its exoskeleton, resting on a reed, drying its body and gauzy wings. In a piece of divine hocus pocus, it will change from boring beige to the black-blue-yellow of the adult common hawker (*Aeshna juncea*) in hours.

If I had arrived an hour earlier to watch the dragonfly emerge from its nymph case, into which it was packed like a telescope, I would have gained 60 *I-Spy* points.

12 JULY: Pink evening glow on the bank's Hereford-shire sandstone. Rose wash. The lifeless scum along the east edge of the pond absorbs the dying sun's rays like blotting paper. A single heron's wing feather lies on the water, a quill waiting to write.

The surface of the pond is natural palimpsest, continually being rewritten, continually recording everything. A single drop of dew off the alder. And, today, the passage through the duckweed of something large enough to leave an open channel. A rat, maybe. Or maybe a rat is too small.

I enter the water trepidly, after a carp jumps up at me; the water is as warm as soup, and as thick. It sticks to me. One circle is enough. On getting out I rub myself with a towel, which greens with each swipe. I am the green man.

The algal scum is thickest under the windbreaks at the pond ends provided by the alder and the briar. Bubbles of methane have been trapped under it; green boils on the face of the pond.

13 JULY: The sounds and sights of the farm pond at twilight:

9.08pm Daubenton's bats under the alder, flying at break-neck (as in 'break one's neck' to watch them Wurlitzer around). The pond is dark and sleepy with night. The colour of a dead PC screen.

9.16pm A 'flip-flop' on the water. Carp jumping. Temperature dropping, and made fan-cold by the swirling bats. The cooling of the air releases the stink of the summer pond. Grasshopper playing their minute maracas.

9.22pm Tawny owl calling. Light almost vanished; only the dimmest reflection of sky and trees on the water. The pond at night is a daily hibernation; a cycle within a cycle.

9.26pm A frog jumps from the bank into the water. Light a citrus candle to keep the biting flies at bay.

9.30pm Wren *ticks* briefly from the briar, then returns to her slumber.

9.31pm Drowned in darkness, all light gone. Only noises now.

9.33pm Fish splash; then from the alder at the other end, noise of a stick breaking; fox coming to water, but she smells me. Last sound; the swish of an animal running through dense summer meadow.

28 JULY: Daily the pond swelters in the beating sun; the oldest newt larvae, having metamorphosed from their aquatic larval phase to efts, begin to emerge from the pond (the exodus continues until August; it then takes the immature newts between two and four years to reach sexual maturity).

The mud at the pond bottom belching vulgar bubbles of gas; a mesh of common gnats above the water, a foot or more deep; *Culex pipiens* takes the blood of birds, not humans.

In the rushes, a grass snake, the first I have seen this year, hunting frogs. Skitty sounds the alarm, clucking quite fiercely at him, the slithering serpent in Eden.

The pond has soft rush (*Juncus effusus*), in the typical clumps; undisturbed it will grow three feet in height. The stems are glossy and cylindrical, and it is the contents of the stem that made the 'rushlights' of yore. As late as the mid-nineteenth century the pith of soft rush, soaked in fat, was the standard artificial light in the cottage. Farmer and MP William Cobbett declared, 'I was bred and brought up mostly by Rush-light, and I do not find that I see less clearly than other people. Candles certainly were not much used in English labourers' dwellings in the days when they had meat dinners and Sunday coats.'

Fancy cottage self-sufficiency? Gilbert White, Rector of Selborne, provided instructions on the rushlight in his 1789 *Natural History of Selbourne*:

The rushes are in best condition in the height of summer, but they may be gathered, so as to serve the purpose well, quite on to Autumn. It would be

218

needless to add that the largest and longest are best. Decayed labourers, women and children, make it their business to procure and prepare them. As soon as they are cut they must be flung into water, and kept there; for otherwise they will dry and shrink, and the peel will not run. At first a person would find it no easy matter to divest a rush of its peel or rind, so as to leave one regular, even rib from top to bottom that may support the pith: but this, like other feats, soon becomes familiar even to children; and we have seen an old woman, stone-blind performing this business with great dispatch, and seldom failing to strip them with the nicest regularity. When these *junci* are thus prepared, they must lie out on the grass to be bleached, and take the dew for some nights, and afterwards dried in the sun.

Some address is required in dipping these rushes in the scalding fat or grease; but this knack also is to be attained by practice. The careful wife of an industrious Hampshire labourer obtains all her fat for nothing; for she saves the scummings of her bacon-pot for this use; and, if the grease abounds with salt, she causes the salt to precipitate to the bottom, by setting the scummings in a warm oven . . . A pound of common grease may be procured for four pence and about six pounds of grease will dip a pound of rushes . . . if men that keep bees will mix a little wax with the grease, it will give it a consistency, and render it more cleanly, and make the rushes burn longer: mutton suet would have the same effect. A good rush, which measured in length two feet four inches and a half, being minuted, burnt only three minutes short of an hour: and a rush still of greater length has been known to burn one hour and a quarter. These rushes give a good clear light . . .

Those doing fine work, such as darning or lace making, would use a globe of water as a lens to produce a spotlight. A lighted rush would be laid on the edge of a chest of drawers at bedtime giving just enough time to undress and get into bed. The edges of old furniture often have burnt little grooves from this domestic task.

In the Second World War the rushlight was revived in some rural areas.

1 AUGUST: Flies sew a net of sound around the farm pond's quintessential mouldy stink.

Lazing beside the pond . . . A pond, as good as

an oasis to me, a drink in nature. And Orwell comes
to mind:

> . . . in a manner of speaking I AM sentimental about
> my childhood – not my own particular childhood,
> but the civilization which I grew up in and which is
> now, I suppose, just about at its last kick. And fishing
> is somehow typical of that civilization. As soon as
> you think of fishing you think of things that don't
> belong to the modern world. The very idea of sitting
> all day under a willow tree beside a quiet pool – and
> being able to find a quiet pool to sit beside – belongs
> to the time before the war, before the radio, before
> aeroplanes, before Hitler. There's a kind of peaceful-
> ness even in the names of English coarse fish. Roach,
> rudd, dace, bleak, barbel, bream, gudgeon, pike,
> chub, carp, tench. They're solid kind of names. The
> people who made them up hadn't heard of machine-
> guns, they didn't live in terror of the sack or spend
> their time eating aspirins, going to the pictures, and
> wondering how to keep out of the concentration
> camp.

I am not really a traveller, although in this book I jour-
neyed more than any other time in my life. I like to

know one thing well; something I can contain and understand.

So, I went night-walking last night around my place.

I'd gone out to check the piglets; eight of them, black and brand new. In the torch's glare, they were packed as tight as sausages on their mother's teats. Exhausted, she barely managed to grunt a greeting.

Piglets are not pretty; they have the ribbiness of the concentration-camp starved and the plunger-mouths of the bourgeois greedy. The straw in the ark lay like heaped strands of shredded sunshine (which in a way it was). Turning to leave the pig paddock for the house, I saw the lane, luminescent and winding into the night.

I thought, why not go for a tramp? It was sultry, or 'close'; the sky was clamping, and painted in heavy mauve oils of stratus cloud. Sleep would have been difficult anyway. And I had a Labrador, Bluebell, with me.

We started along the open lane, the dog and I, heading west. We were wandering free in a new land: The Night. From a gatepost a bobby little owl squeaked, mimicking polystyrene being twisted. This was eleven thirty or so, beyond the time when humans are about.

Against the pale of the horizon the oaks were as exact as razor-cut silhouettes.

In the cornfield, the ziggurats of bales awaited collection. The harvest scene was hushed and breathless: something from a black-and-white, older England.

We stood by the open entrance: in the shades of greyscale the badger in the stubble was detectable only by its shuffling gait, which is that of an old man in baggy trousers. I supposed *Meles meles* was scavenging for carrion caused by the revolving steel blades of the corn-cutter.

We left Mr Teddy, as we call badger in Herefordshire, to his macabre meal. The dog's panting counted cadence as we headed uphill; there was so little human noise at that time of night that the traffic on the A465, more than two miles away, assumed the musical murmur of a stream.

It is the bane of modern Britain, is it not, traffic din?

From a farm across the wooded valley, a collie barked, and the sound bounded across the dry earth, as though a thrown ball.

Going down the hill, under the clotted leaves of the arching lane-side oaks, we were submerged in black shadow, so there was no up or down, left or right. Just the original darkness of the universe. All around was the leather-whisper of bats' wings. For a moment I turned the torch on; the air whirled with a snow storm of moths. One pursuing pipistrelle flew close enough for me to feel the breath of its wing on my face. It was as cooling as a fan.

August is the month of peak bat; it is the month when young bats leave the breast of the dam.

Emerging from the oaks, we encountered the farm pond staring at the sky; beside the pond lay ten Simmental cows, sculpted in stone. The hedge was cut low, and I moved closer to have a good look. I like cows. Few things smell as lovely as pasture-fed cows; they wear the perfume of buttercups and clover and cut grass.

Then I realized there was an animal in the pond, swimming with purpose, making great divergent lines behind it. For a moment I thought it was an otter; then realized the unlikeliness of that.

As if to answer my wondering, the fox walked out of the water and shook itself. Why was it night-swimming? To catch some poor pond creature, I suppose. My grandfather, who hunted and farmed and knew foxes like kin, told me the 'cunningest' thing he had ever seen a fox do was walk along the top of a freshly cut hedge, making use of each new shrub-cut to place a single paw. By this intricate balancing act it walked to the hen yard. And in.

From tonight's uncut, sprawling hedge, two small

birds fluttered out, making a childlike complaint about my gawping; I felt guilty, and wished them a safe roost.

The last summer month is all change in the kingdom of the English mammal; hedgehogs have a second brood, while the first go off into the wide world. A young hedgehog tippy-tottered across the lane; it was on its own personal night-walking adventure.

The lights of Hereford and South Wales were far off, orange smears on the horizon.

On we went. A deserted barn loomed like death's head. From its bald, slate roof floated up a silent predatorial orb: a tawny owl. Down by the copse there was the wail of a rabbit losing its life, perhaps to the owl's kin, the sound of the English night for a thousand years.

Suddenly, from nowhere, a car came along the lane, blaring light and rap music. I turned on the pocket torch to make us visible in the darkness. The car's horn shrieked, louder than the scream of the dying rabbit. Someone shouted an obscenity through the open windows; the raised voices trailed behind with the exhaust fumes.

Apparently, no one has the right to walk the roads at night in the time of the car.

It was hot, we began to fade.

The dog and I took a single-track lane off to the

left, to make a circle in the arterial map of the old byways that serve the countryside. We were lost in solitude again.

The dog went on ahead, around a bend. I called her back, shone the torch, which picked out her green eyes. It was then that I realized that Bluebell was behind me, down in the ditch. The emerald jewels belonged to a she-fox. She was wet, her fur pasted close. I saw then that she was the swimmer in my pond.

The vixen showed no fear, standing there on the lane. She stared back.

Like the people in the car, she seemed to wonder by what right I thought I could wander the lanes at night. Afterwards, I, a trespasser from daylight, went to earth.

4 AUGUST: All the textbooks insist that adult frogs leave the pond by mid-July; yet here are three sitting in the weed of the farm pond, eyeing me. They are draped with green vegetation, it lies on them like noble robes.

Today the mallard and her sole surviving offspring (all grown up) reappeared from who knows where. They are back. What do I care if they treat the place like a hotel?

*

21 AUGUST: We cut the hay in the meadow a fortnight ago, leaving a three-yard boundary around the
pond; from above it must look like a monk's tonsure.

Today we let in the cows. They run, tails up,
straight to the pond, trampling the mint and burdock,
to drink; the water drips from their great pink mouths
when they lift their heads.

The pictures of satisfaction are the cat that got the
cream, and the cows that got the pond water.

The cow pond of the old England was killed by the
tap. When no longer used by farm animals – which
trampled the weeds down like our cattle are doing
today – the weeds took over.

There were other reasons back in the good old
days to maintain the farmland pond, and tame the
ever-invading trees and scrub: fishing by villagers;
removing the natural home of the brown rat; providing firewood.

Today, many of the rural reasons for controlling
pond 'succession' no longer apply.

No more pond-swimming for me. Cows are wonderful;
their effluvia in water is not, and livestock excrement
in water is one of the reasons farmworkers are particularly prone to Weil's disease.

*

On an evening walk up the lane, I poke about in a short, fenny section of ditch carrying water still, almost buried beneath tumbled-over hemlock in seed. A frog jumps out at me.

Only in the moment of its drying up is a ditch a linear pond. When water courses through a ditch it is a stream.

26 AUGUST: I've come to think of it as 'The Walled Garden Mystery'.

For three nights now, rabbits have been gnawing their way through the salad patch, until there is scarcely a leaf of radicchio left. (They seem, incidentally, immune to the attraction of rocket.)

The conundrum? How the rabbits get in. The brick walls are old but six feet high, the pedestrian door is varnish-blistered but solid plank, and the galvanized field gate, the obvious point of entry, sheeted in anti-rabbit wire mesh. Yesterday, I patrolled the outside walls, looking for tunnels in the manner of an especially diligent Colditz guard. For one tired moment I paranoically imagined genetic-freak rabbits hopping over the walls.

This morning as I surveyed the bone-stumps of lettuce, I realized I was in a horror story not a detective howdunnit, a garden-time Stephen King not an Agatha Christie. The rabbits were inside the walled garden. I had locked them in on Tuesday evening after toing and froing through the field gate with the wheelbarrow when lifting the potatoes.

Sure enough, when I examined the 'wildlife area', this contained more fauna than intended.

Such is our hubris: we want nature, yet only where and when we want it.

The wildlife garden, a ten-foot strip along the bottom wall, has the usual accoutrements to attract Mrs Tiggy-Winkle and Buzzy Bee: an hibernaculum consisting of a half-buried sherry cask and 'hotel' of racked hollow hogweed stems, respectively. There is a bat box, a bird box. The butterflies have a buddleia. For Gussie Fink-Nottle's newts it is the land of milk and honey and des res: mini-pond (a disused water trough with a ladder of stones and filled with water milfoil), rock pile, log pile and long grass.

The unintended feature, deep under the briar, is an explosion of salmon earth, as if a rocket had ploughed into the ground. A fresh rabbit burrow.

The rabbit, like central heating and apples, came over with the Romans; it has been nibbling its way through Mr McGregor's garden produce – and mine – ever since.

I thought about shooting the intruders, but if there were more than one I'd need a machine gun, or to devote hours to sniping with the .22 Weihrauch. Besides, what would I do with the flesh? I like rabbit in the casserole pot in winter, with a spoonful of Dijon mustard and a handful of tarragon; but in summer it is too gamey.

Coney, bunny, call it what you will, rabbit is sniffily overlooked in twenty-first-century Britain. We were not so fussy in 1939–45, when rationing was the order of the day; then rabbit kept the nation's stomach from rumbling. Rabbit is lean meat, but healthy meat, replete with selenium, iron and zinc. But in summer, as I say, it is too gamey.

So to rid the walled garden of the rabbits, I have decided to hide behind the compost bins, and shoo them out through the opened field gate when they emerge from the burrow . . .

The evening sun is a ball of red. A swallow lands on the telephone wire, and her throat patch and the sun are a match.

The brick walls make still and make silent: a cloistered peace upon the geometric rows and beds of vegetables, herbs and fruit trees. (It is only a small walled garden: thirty yards by thirty yards.) There is just the slight ticking of contracting, two-hundred-year-old brick as it cools, and on the wall a solo blackbird singing evensong, sad and spiritual. This late in the summer, the blackbird is the only garden songster.

The heat of the evening is sultry and panting-hot, as befits a 'dog-day', named for Sirius the Dog Star, though to my mind for the type of heat that leaves Labradors gape-mouthed. In the rose-glow, the aroma of rosemary and thyme rises in a scented tide.

I wait, and I wait in the trapped stillness and miniature scenes.

Beside me, a toad emerges from the earth. He waddles across the grass path, ungainly, Winston Churchill in chafing shorts, into a tipi of runner beans, to give a lascivious flick of his tongue at some morsel.

I pick him up; he is unpleasant to hold; a cold handshake. He 'perfumes' my hand by releasing milky, repellent bufotoxins from the warts on his neck.

The toad has come from the farm pond; in a month or more, he has walked eight hundred yards. He will stay in the garden until next year, or, if he is luckless, until he is caught. No successful pond is an island; the pond and its inhabitants need the wider environment too.

Distracted by the toad, I fail to notice, through the slats of the compost bin, the briar rabbits emerging. One is already on the path; behind, in the shaggy grass, are the tips of four pairs of pointy ears.

The lead rabbit, a doe, is cautious; a shadow-shape, she rises on her haunches to better smell the night air. After a minute of nose-twitching, gluttony overcomes discretion and she proceeds towards the salad remnants. Hoppity, hoppity. Her young follow her. Hoppity, hoppity.

I give the rabbits two minutes to commence eating. Then I dash out, to get between them and their safe burrow home.

The rabbits are gratifyingly surprised to see a middle-aged farmer appear behind them, shouting 'Yah! Yah!' (*A priori* I decided that my cow-moving command was appropriate to the endeavour.)

Rabbits have no flock instinct. The five of them separately pin-ball around the walls, the water butt, the fruit trees, the shed, until eventually they all bounce out through the field gate.

I watch them run down the field, their white scuts bobbing in the last light, towards the wood.

Where I think they belong.

Untitled

The old pond full of flags and fenced around
With trees and bushes trailing to the ground

The water weeds are all around the brink
And one clear place where cattle go to drink
From year to year the schoolboy thither steals
And muddys round the place to catch the eels
The cowboy often hiding from the flies
Lies there and plaits the rushcap as he lies
The hissing owl sits moping all the day
And hears his song and never flies away
The pink nest hangs up upon the branch so thin
The young ones caw and seem as tumbling in
While around them thrums the purple dragon flye
And great white butterflye goes dancing by.

John Clare

THE POND IN AUTUMN

La Baignade, Nueil-les-Aubiers, Western France
1 September: Nueil-les-Aubiers is one of the poorest towns in Western France. As if to make amends, successive mayors have prettified the place with nature. Down in the park beside the Scie, there is a new open-air *baignade*, a natural swimming pool pleasantly situated in a sculpted landscape of grass and trees; one would hardly know that the back of Intermarché was fifty yards away. The white spire of St Hilaire peers bucolically over old scrub oak. It could be deep country, not a town suburb.

The *baignade* – which boasts a chute, and a cafe with decking – is barely fenced off; the surrounding wire-netting is four feet high. In Britain, everyone would be climbing over for free entry; the French are different from you and me. They, rule-abiding to the nth degree, pay.

Nudged to behave well, we hand over the five euros admission each. It is one of those bonus days when it still seems high summer, rather than its last call.

There are more than a hundred people at the *baignade*, in the water or sprawled on the banks, mostly under eighteen. There is none of the out-of-control semi-violent fairground atmosphere you might get at a normal pool; it is as if everyone is calmed by nature.

The Nueil *baignade* is no normal swimming pool.

It is 'biologique'. The main area is green plastic-covered concrete, but the ends are reeds and gravel, and it is these that cleanse the water. One of the reed beds is five yards by fifty yards, and would not disgrace a fen. Speckling the pallisade of reeds are yellow iris and water soldier (*Stratiotes aloides*), the first nostalgically familiar, the latter entirely foreign to me; its rosette of long leaves are tooth-edged in a dandelion sort of way, yet firmer. They could cut flesh.

There is no chlorine in the *baignade* whatsoever.

Ergo: I like to think of Nueil as a pond.

The water is palest green (there is a light dusting of algae on the sides), crystal and cold; the sort of purity of water you expect to find only in mountain streams. I swim with a freedom in pond water I don't usually find, since the submarine life forms I meet are human, rather than crucian carp. There is pleasure in carving the lucid water, with precise artisanal strokes.

It is 26°C even at 7pm. The absence of chemicals means that when I climb the metal ladder out I feel refreshed, and my skin has none of the emaciated whiteness that comes from thirty minutes in chlorine. Ultimately chlorine kills – the fun of swimming, as much as the intended bugs.

Is Nueil really a pond? There are southern damsel-flies (*Coenagrion mercuriale*) and emerald damselflies

(*Lestes sponsa*) breeding in the reeds, and they drift over the swimming area.

Well, I've never seen a swimming pool with a damselfly at home.

The Farm Pond, Herefordshire
5 SEPTEMBER: Sitting beside the pond, I have an indescribable sense of panic, as though nature had become malevolent. I scoot away, out in the meadow, familiar and dry ground. Something similar once happened to me in a wood, where I thought the trees were closing in, gathering around. I ran then too.

Later, ashamed, I go back, to watch the Daubenton's bats, which are coming out earlier and earlier, to skim over the pond in search of insects. It is then I see the grass snake, swimming just under the surface of the water.

The Mill Pond, Argenton, Western France
8 SEPTEMBER: A red dragonfly (the same red as burning blood), which requires consultation with Collins *Freshwater Life*, now kept in the tool box of the tractor. A 'ruddy darter', but perhaps 'bloody darter' would be more apposite since its scientific name is *Sympetrum sanguineum*.

The dragonflies will not fly for much longer; we have almost reached the last *danse*.

Bluebell leaps in for a swim, her tail twirling around like a propeller. I wait on the bank; the air is so still that even the aspen who wait to agitate are motionless.

The willow bark runs and twists, perpetually perplexing the treecreeper.

The water is at its lowest level, so I can see the pond bottom in its entirety if I walk around the edge. Over everything – the rocks, the sticks, the leaves, the skeleton of a rat – a layer of antique yellow silt. It is like looking at the contents of a tomb, from behind glass.

The Farm Pond, Herefordshire
20 SEPTEMBER: Drakes leading the way in their new finery in the resumed courtship circle; the alder's leaves moth-eaten. Two seasons in one scene. The year turning full circle.

The pond; one can almost grasp it with your hand, and always that tincture of rot which is simultaneously life-giving; the pond is the primordial ooze right here, right now, unlike the vast and saline sea, which always has the feeling of sterility.

Ponds should not be more than 80m across: the point at which they stop 'stratifying', or developing different ecological zones – very shallow, shallow, deep, very deep. A crucial difference between pond and sea is that most of the bottom of the latter is samey.

25 SEPTEMBER: Today I have my own private rush-cutting ceremony. In the past of England, bulrushes were cut – normally in June or July when the ruddy-brown flower spikes are in bloom – for weaving, paper and covering the floor before the age of the carpet. The household roll of Edward II (1307–27) shows a payment to a John de Carlford for 'supply of rushes for strewing the Kings chamber'. In churches the strewing of rushes, for cleanliness and insulation, was done at least once a year, and was eventually turned into the Rushbearing Festival, complete with all sorts of jollities. *The History of the County of Derby* (1829) gives descriptions of the rushbearing at Glossop:

> Previously to our leaving Glossop we visited the village church . . . Here we observed the remains of some garlands hung up near to the entrance into the chancel. They were mementos of a custom of a rather singular nature, that lingers about this part of

Derbyshire, after having been lost in nearly every other. It is denominated rush-bearing; and the ceremonies of this truly rural fête take place annually, on one of the days appropriated to the wake or village festival. A car or wagon is on this occasion decorated with rushes. A pyramid of rushes, ornamented with wreaths of flowers, and surmounted with a garland, occupies the centre of the car, which is usually bestrewed with the choicest flowers that the meadows of Glossop Dale can produce, and liberally furnished with flags and streamers. Thus prepared, it is drawn through the different parts of the village, preceded by groups of dancers and a band of music. All the ribands in the place may be said to be in requisition on this festive day, and he who is the greatest favourite amongst the lasses is generally the gayest personage in the cavalcade. After parading the village, the car stops at the church gates, where it is dismantled of its honours. The rushes and flowers are then taken into the church and strewed amongst the pews and along the floors, and the garlands are hung up near the entrance into the chancel, in remembrance of the day. The ceremony being ended, the various parties who made up the procession retire, amidst music and dancing, to the village inn, where they spend the remainder of the day in joyous festivity.

Rushbearing often attracted unsavoury characters, such as cutpurses and pickpockets, and became a pretext for heavy drinking, which led to intemperance and indecorum, even among pillars of the community:

> Tristram Tyldedesly, the minister at Rufford and Marsden on Sundays and hollidaies hath danced emongst light and youthful companie both men and women at weddings, drynkings and rushbearings; and in his dancing and after wantonlye and dissolutely he kissed a mayd . . . whereat divers persons were offended and so sore grieved that there was weapons drawn and great dissenssion arose.

During the nineteenth century the festival died out in most parts of the country, for the simple reason that church floors were laid down to stone. The tradition, in a spirit of rustic Romanticism (or community building), has been revived in the north. Such as at Sowerby Bridge:

> Over the course of the weekend, our own festival sees the progress of the rushbearing procession around 7 towns and villages visiting many churches and local hostelries along the way. The focal point of the procession is the sixteen feet high, two-wheeled,

handsomely decorated and thatched rushcart. A team of young ladies take turns to ride on top of the cart as it is pulled by sixty local men dressed in Panama hats, white shirts, black trousers and clogs. Accompanying them are a group of supporters in Edwardian dress along with some of the region's finest musicians and Morris dancing teams to provide entertainment for the crowds.

Anyway, I am rush-cutting to save a stupid cow that has gone too far into the pond, and has decided it cannot get out of the sucking mud, and has spent the morning trumpeting, loud enough to wake neighbours a mile away, and the rest of the herd has gathered mooing around the edge of the pond, spectators at theatre in the round.

I need to tow her out with the tractor, a chain around her body. But the chain, the one with links the size of fists, is only twenty-five feet long; meaning, in short, for the chain to work, I have to reverse the tractor on to the cows' access ramp into the pond.

And there's the rub. The ramp is muddy. Slippery. Soft. So for the tractor wheels to purchase, the ramp needs to be something thick and grippy.

I confess it takes me a while. Sacks? Wire netting? Rocks?

How now, brown cow, do I get you out?

I am staring at the silly moo in the pond, then realize I am staring at the solution. A mat of bulrushes. So off I pop to the barn where at the cobwebby rear I keep the museum of agricultural curiosities. 'Why do you keep all that old junk?' farmers are asked.

Well, for moments like this: I find the Edwardian billhook, as heavy and satisfying in the hand as butcher's cleaver.

A quick sharpen, rub thumb laterally over the blade – could shave with it! – and back across to the pond, the sun shining, the robin singing a happy tune rather than the usual theosophy. Get in tractor, drive with billhook to pond; from across the meadow, with the cows grouped around, this can pass for an oasis in the African savannah.

Jump down, edge into the bulrushes, which must be eight feet high, two inches in circumference around the base, before they fan into a display of green swords.

Slash with the flashing steel. Satisfying liquid noise, as when cucumber is broken. The bulrush spire tumbles.

In ten minutes of old-school bulrush cutting I have enough rushes for a mat on the ramp six inches deep.

Reverse tractor. Wade (waist high) to cow, with chain. She gives me the 'What took you so long?' look. Fix chain behind over her back and under her front legs; attach chain to tow bar of tractor.

Climb up into cab. Select second gear (bit more traction than first), pull the throttle round for some power. Big diesel plumes. (Not to be taken into Central London, our International.) Inch forward. Look behind, as chain-slack taken up. The cow is already moving, lifting up her front legs, wallowing forward.

The rushes hold. A bit more power. Then she's on the ramp, and out.

It takes a week for the pink mud in the pond to settle. This is about the same amount of time as it takes me to clean out the mud from the tractor cab.

I will not name and shame the cow who got stuck.

Humans domesticated *Bos taurus* around 7000 BC, in the Middle East. They have been getting into trouble down on the farm ever since. Our neighbour Lindsay Lloyd once had a cow who walked over the edge of the quarry, and fell thirty feet into mire, and had to be pulled out by the Ewyas Harold part-time fire brigade. Pure *Trumpton*.

*

27 SEPTEMBER: A swallow flies low over pond, takes a sip on the wing, before heading south for Africa; the moorhen remains behind.

Mirabelle the cow still looks as though she has been lifted up and dipped in pink dye up to her middle.

In the alder, in the heat of the afternoon, a pigeon *roo coos* intermittently. It is always old England under the soft, absent-minded cooing of the wood pigeon.

28 SEPTEMBER: One of those bright cold autumn mornings when the pond is a perfect natural mirror, reflecting the sky and the trees with complete purity. It is a mirror that can never be broken; sometimes it is dulled by errant cows and pollen dust, tarnished by algae, misted by the breath of fog, but the wind and the rain always refresh it, eventually.

Skitty's last chick, all grown up, has gone away. I hope to a pond of her own. She looked the spitting image of her mother; only the olive in her wing coverts viewed through binoculars showed her young age. A mature moorhen has slate-black wing coverts.

The Mill Pond, Argenton, Western France
29 SEPTEMBER: Kingfisher on the volcanic rock on the pond, turns to display his back, the vertical fluorescent stripe of gas-light blue, a boxer in a cape; the long bill particularly noticeable in profile. A harpoon.

Kingfisher: the bird's alternative name, Halcyon, speaks of Greek myth; Alcyone, the moon goddess associated with the solstice, tried to drown herself after Zeus killed her husband Ceyx, and was turned into a kingfisher.

In D. H. Lawrence's *The Rainbow*, Ursula Brangwen escapes her parents for the first time; she walks through the woods until she comes to a river, she sees a kingfisher darting blue – and then she is transfigured into happiness. The kingfisher is the witness of the order of enchantment.

As Kingfishers Catch Fire

As kingfishers catch fire, dragonflies draw flame;
As tumbled over rim in roundy wells
Stones ring; like each tucked string tells, each hung bell's
Bow swung finds tongue to fling out broad its name;
Each mortal thing does one thing and the same:
Deals out that being indoors each one dwells;

Selves – goes itself; myself it speaks and spells,
Crying Whát I dó is me: for that I came.

I say móre: the just man justices;
Keeps grace: thát keeps all his goings graces;
Acts in God's eye what in God's eye he is –
Christ – for Christ plays in ten thousand places,
Lovely in limbs, and lovely in eyes not his
To the Father through the features of men's faces.
                                                    Gerard Manley Hopkins

Leaves on the pond in a flotilla, like the craft on the Thames for Handel's water music.

First showing of daddy long legs.

On the other side of the Argenton, hidden by the islands, a car stops, a door opens and slams. The dread sound of a fisherman arriving to break the peace. The French are mad about fishing; even a local *magasin de journaux* will have a fishing section, along with a display of wine.

The fisherman sets up, throws sweetcorn bait in the water, flicks out an orange float.

And the ripples he makes: the sunning otter has long gone.

*

30 SEPTEMBER: Meadow brown butterflies settle on the nettles beside the mill pond, before flying haphazardly away.

Tree frogs clacking in the wisteria; the common frogs as still as the pond stones on which they sit.

I look away, and they go away.

Sunlight illuminating the bottom; a coypu detaches the hawser of a water lily, and drags the plant on to the bank to nibble.

I watch a wren in the willow, and realize that the classification of animals, the taxonomy, is wrong. It matters not a jot whether the wren is a bird or a mouse.

There are only the hunter and the hunted. A wren is mouse with wings.

Also in the willow, a band of long-tailed tits. You never see a long-tailed tit by itself.

The surprising violence of falling willow leaves; they strike the surface coat of the pond petulantly in playground blows.

Standing on the stone bridge, admiring the slow Argenton: see the kingfisher from above; he comes through the central span at speed, 'planing' the river – always keeping the same height, in the manner of magnetic train on a magnetic track.

Is it a bird? Is it a plane? An equation? A physical state. The river and the bird live in a state of affinity; the river is the kingfisher's road. Bird and river belong together.

I have never seen a kingfisher fly over a field, through a wood.

He perches on the ruined mill wall, fire-chested; then flies off, the sky on his back.

The Farm Pond, Herefordshire
2 OCTOBER: All the rushes and reeds bowed over; the alder half leafless and enwrapped by rain. On the cow's access ramp, piles of beetle shells; larvae that have pupated in the mud through summer have emerged as adults in a 'mass emergency'. Predators know about collective insect birthings, and binge on them. Reading the hieroglyphics in the mud, I see that the heron, fox, vole (tiny claws) have all attended the banquet.

9 OCTOBER: Watching the ripples dash over on the silver surface of the pond, in ribs of water; the pond always betrays the wind.

I've not seen a pond skater for a week. The frost has put paid to their antics, until next year. On calm days now only the wind, the dropping leaves and Skitty disturb the surface of the waters of my piece of earth.

Skitty has lived on the pond for two years now. It belongs to her.

At night: the stars are diamonds spread across black velvet.

10 OCTOBER: In the September of 1921, sixty-six-year-old Alfred Watkins delivered a paper entitled 'Early British Trackways, Moats, Mounds, Camps, and Sites' to the Woolhope Naturalists' Field Club (formed in 1851) in Hereford, presenting his thesis that the landscape could be mapped as a series of straight lines or 'alignments' that 'connected ancient burial mounds, monuments, barrows, ditches, castles, ponds and trackways'. Watkins went on to name such alignments 'ley lines', and attributed their causation to the navigations of early man.

In rapid succession Watkins published *Early British Trackways, Archaic Tracks around Cambridge,* and then, in 1925, his magnum opus, *The Old Straight Track.*

Watkins belonged to the twilight of the generalist

and the amateur, those private Victorian and Edwardian researchers who felt no need to stick to narrow specialisms, but let their minds wander over wide bodies of knowledge, confident that all that could be known on a particular topic was readily available to them.

Born at the Imperial Hotel in Widemarsh Street, Hereford, on 27 January 1855, Alfred Watkins was the son of a Victorian entrepreneur who expanded from owning the Imperial Hotel to owning the Bewell Street Hereford Brewery and The Friar Street Flour Mill.

The brewery was sold in 1898 for £64,000, making Alfred Watkins totally independent and able to indulge his hobbies. He had great success inventing the 'bee meter', a device to calculate photography exposure times (it was used in Robert Falcon Scott's 1910 Antarctic expedition), and he became a fellow of the Royal Photographic Society.

His other hobby was scouring England for ley lines. Although he created a network of ley hunters, the Straight Track Club, to expand on his work, Watkins was never able to convince professional archaeologists of merits of his theory.

But *The Old Straight Track* got an unexpected relaunch in 1969 with the appearance of John Mitchell's

The View Over Atlantis. Mitchell converted Watkins' ley lines into a sort of mystical national grid, lines of invisible earth energy linking the religious sites of pre-historic Britain.

Suddenly, the conservative miller from Hereford morphed into a countercultural sage.

Unfortunately, Watkin's theory of ley lines holds as much water as a colander. One of his 'sighting points' is the pond, about which he wrote:

> I have suggested how these might have developed from the tump [mound or hill], and shown where pond and tump were used together. Moats are a similar arrangement on a larger scale. The trackways go straight for the island part of the moat. It is not the least amazing part of this revelation that I find practically all the small horse or cattle ponds in field or homestead which are marked on a 6in ordnance map have leys running through them, and that examination in dry seasons shows signs of the road passing through them. 'And when we cleaned the pond out we found it cobbled at the bottom' is a frequent report made by a farmer . . . I cannot say that passengers walked through the bottom of these ponds (most of them have one shelving edge, with the opposite bank steep), but

to this day an ancient road (at Harley Court, Hereford) does go through the bottom of a small pond . . .

I've tested this theory to destruction. Take an Ordnance Survey map, and you can find straightish lines between a pond and two or three other landmarks without trying. How could you not when Watkins assigned any landmark feature a role as signpost? Ponds had shelving edges but this was so carts could be driven to hydrate the wooden wheels, and cattle could enter to drink, sheep be washed.

I am a fan of the pond. But even I cannot be persuaded that, among its wonders, it is a signpost for travellers.

My stepmother, my ancient and wonderful stepmother, lives in a retirement flat in the named-for-him Watkins Court, Hereford . . . A princess in some time-inverted fairy tale.

18 OCTOBER: The rain has put a foot of water on the pond, so it is almost up to the rim, and what mud is still left exposed has absorbed moisture to become as squidgy as sponge. In this soft state, every rain drop falling from alder makes a dimple.

A rope of vixen scent around the alder and the briars; she is hunting Skitty, but Skitty is keeping close to the briars. I toss her a handful of corn every day.

She is the one true pond being; she is always here. The others they come and they go. Skitty is the bustling busy house-keeper in the background. She is the pond-keeper.

A heron comes in, ancient and slow; at the last moment it notices me standing by the alder trunk, and veers away looking abject in the way that herons do, as though life is too much trouble.

High Overhead That Silent Throne

High overhead that silent throne
Of wild and cloud betravelled sky
That makes ones loneliness more lone
Sends forth a crank and reedy cry
I look the crane [heron] is sailing oer
That pathless world without a mate
The heath looked brown and dull before
But now tis more then desolate

John Clare

*

The Farm Pond, Herefordshire

22 OCTOBER: An overnight rainstorm fills the pond, satisfying it, making it replete. After the downpour the pond is as still as I have ever seen it, but the run-off from the field has turned it the pink of blancmange.

Three mallard wing-whirr away before I get within a hundred yards. They launch into the wind. For the lift, you see?

Then, one more mallard, a female, breaks late and quacky out of the reeds. In her ascension I catch her familiar eye.

26 OCTOBER: In the Sevylor, in the middle of the pond, in the middle of mist. The undulations from my rowing hitting the reeds, and returning; the slap of water on the inflatable canoe the only sound in the world.

I seem to be floating in mid-air.

In *Walden* Thoreau once took to the water and remarked that the surface, 'being so smooth, betrayed where a spring welled up from the bottom'. I realized yesterday, I do not know where on the pond bottom the spring is. So I thought I would find out.

Try as I might, I can see no super-smooth patch, so I change technique, and lie back, and paddle around with my hands.

By Skitty's briar patch, there is a definite temperature difference, it a degree or two colder on a hand already freezing; and if I stick in my arm (shirt sleeve rolled up) I can detect the slightest upwards thrust of water.

Of course, Skitty has made her home by the spring. The rising of the water disturbs the pond bottom, bringing mites and morsels to the surface. Home delivery, nature's way.

The Mill Pond, Argenton, Western France
26 OCTOBER: A red darter (*Sympetrum striolatum*), demented, probing the breeze, again and again, as if raging against the coming of winter.

The Farm Pond, Herefordshire
29 OCTOBER: Pale pallid light; a draped veil of cobweb strung between the bayonets of yellow flag. The webs have the pattern of smashed glass. With the coming of autumn the pace of life is slowing, the water developing its winter gait.

Not much happens at the pond today. It is just Skitty and me.

*

At Rushy-Pond

On the frigid face of the heath-hemmed pond
There shaped the half-grown moon:
Winged whiffs from the north with a husky croon
Blew over and beyond.

And the wind flapped the moon in its float on the pool,
And stretched it to oval form;
Then corkscrewed it like a wriggling worm;
Then wanned it weariful.

And I cared not for conning the sky above
Where hung the substant thing,
For my thought was earthward sojourning
On the scene I had vision of.

Since there it was once, in a secret year,
I had called a woman to me
From across this water, ardently –
And practised to keep her near;

Till the last weak love-words had been said,
And ended was her time,
And blurred the bloomage of her prime,
And white the earlier red.

And the troubled orb in the pond's sad shine
Was her very wraith, as scanned
When she withdrew thence, mirrored, and
Her days dropped out of mine.

Thomas Hardy

The Farm Pond, Herefordshire
31 OCTOBER: Late morning, and I arrive at the pond
with the Kubota mini-digger; it is on caterpillar tracks,
so it sounds like a tank. I apologize to Skitty for the
noise.

The wind is norse-ripping alder leaves, which fall
on the pond, starved and wild. The reeds and rushes
have collapsed with the weight of the season, the passing of time. The pond is falling in on itself, and needs
the helping salvationist hand of humanity with a
Kubota.

Before I start dredging, I take a moment to revere
the water. As D. H. Lawrence once noted: 'Water is
H_2O, hydrogen two parts, oxygen one, but there is also
a third – that thing that makes it water. And nobody
knows what that is.'

Even the basic element of the pond is a divine
mystery.

Epilogue:
The Garden Pond

Ponds are magnets for wildlife. Of all the habitats that one can create to aid nature a pond is the best. Any pond is better than no pond, and even a sunken bucket will make a difference to the number and types of animals that visit a garden. Dragonflies are quite content breeding in old washbasins and Mrs Tiggy-Winkle will happily lap from an upturned metal dustbin lid. But as a general rule the bigger the pond the better: the big pond provides a range of complex mini-habitats, and deep water is less likely to boil in high summer, freeze solid in winter. A hibernating frog will thank you for a deep end of about 60–90cm.

Some of the loss of ponds in the countryside has been offset by the suburban gardener, roughly one in ten British back gardens now having a pond. A pond in the garden is an individual, incremental increase in the freshwater habitat; a series of ponds in a neighbourhood can establish an aquatic 'corridor' that allows wildlife to move around.

Ponds also store carbon, helping to reduce global warming.

Ponds, aside from providing for resident wildlife, are watering holes for passing birds and mammals, from migrating redwings to the neighbourhood fox on her nightly perambulation.

Ponds provide endless interest and drama for humans.

Ponds are a place of tranquillity and beauty in the garden. A place for reflection.

Ponds. What's not to like?

Step-by-step: How to Build a Wildlife Pond in the Garden
Depending on size and type, a pond will take a weekend or two to complete. It's best to dig a pond in autumn when the ground is soft, and let it fill naturally with rainwater. If you are installing a pond at any other time of year try and store rainwater beforehand to fill it – tap water is death to ponds because it often has high levels of nitrogen and phosphorus, which encourage 'green pond syndrome', an attack of choking algae.

Siting
Where you dig your pond is critical, because a shallow pond in a very sunny site will be prone to algal bloom, drying out, boiling up. Shallow water is heated more efficiently by the sun than deeper water and, in combination with high nutrient levels in the water, is also a recipe for problem weed growth. Part shade, part sun is perfect – and this can be easily manufactured by

planting tall marginal plants (plants of the edge or 'margin') such as yellow flag. Tadpoles adore bathing in shallow, well-lit warm water; as adult frogs they prefer lurking in moist shade. And you will want frogs in your pond.

Remember that, when established, the pond may need topping up with water; the easiest way to accomplish this is to connect a hose to the overflow of a water butt and, as if by magic, the pond will fill automatically during rainy days. So for the ecological gardener – or maybe the lazy gardener – the choicest position for the pond is within reach of a natural water supply. Conversely, do not position a pond where it will receive run-off from a road or a chemically laden lawn.

Choosing a liner

Unless you live in an area of heavy clay, your pond will need a liner. For small ponds a glass fibre or moulded pool can be used. Their disadvantage is the limited range of shapes on offer, and their tendency to have sides too steep for wildlife.

A flexible liner of butyl rubber or heavy-duty stabilized PVC is the other option; received opinion always plumps for butyl rubber but the latest heavy-duty stabilized PVC is comparable. And cheaper. The great advantage of a flexible liner is that the pond can be architected to suit your and your wildlife's dreams. Flexible liners, however, are fiddlier to fit.

What size and shape liner? Between four and five square metres is the minimum for a pond that combines deep water and shallows. At least two sides of the pond should be gently sloping up to finish in a broad shallow zone. The greatest variety of plants and animals live in shallow water at the edge, often of just 2–5cm depth. Shallows also provide a bathing area for birds.

When designing your own pond remember that a curved or kidney shape looks natural, and is easier to fit a flexible liner into than is a shape with angles.

Vary the side 'profile' by creating ledges of about 25cm in width; these shelves should be 10–20cm under the eventual surface of the water. Alternatively you can create ledges later by stacking rocks and stones.

The deepest part of the pond need not be at the end; the centre is fine. A depth of 90cm will do more than lure male frogs for hibernation, it creates ideal swimming conditions for great crested newts.

To calculate the amount of pond liner you will need for your pond, use this equation:

(Length + 2 depths + 30cm) by
(width + 2 depths + 30cm)

Example: for a pond with the final dimensions of 5m long, 2.5m wide and 0.5m deep:

$$(5 + (2 \times 0.5) + 0.3) \text{ by } (2.5 + (2 \times 0.5) + 0.3) =$$
liner dimensions of 6.3m x 3.8m

The equation allows for an excess to anchor the liner around the edge of the pond.

Excavation
Mark out the shape of the pond with a hose or rope. With a pre-fab pond, simply invert it in the right location and 'trace' around the edge with the hose.

Unless you are very fit or have willing helpers, hire a mini-digger at this point. Whether using machinery or manpower, dig the deepest areas first before establishing the slopes and planting shelves.

Use the barrowloads of topsoil you are suddenly presented with to make a bank behind the pond or to create a raised bed elsewhere in the garden.

Fitting the liner
To avoid puncturing the liner, pre-line the bottom and sides of your pond with unwanted woollen blankets or 3cm of fine builder's sand.

Choose a windless day for fitting, and have a large

pair of scissors handy. Take the liner to one end, and unroll it across the pond, making sure there is excess all the way round. Don't drag the liner when fitting, as any overlooked protruding stones will rip it. Put some bricks on each corner to weigh the liner down.

This is also the time to build ledges from stone and rock, if you haven't 'profiled' these in. To provide an extra cushion under the stones, use hessian sacks or cut spare liner in the shape of the ledge. Put this in position and pile up to the desired height with large stones.

Ideally, you will now fill your pond with rainwater from a butt or two, or it will pour down with rain on cue. Whatever water you add, it will press the liner into shape. Smooth out creases and pleats as they arise.

Once the pond is nearly full, cut the edge of the liner leaving a 30cm overlap all round. To bury the edge of the liner, either: cut horizontally into the surrounding turf and tuck the liner underneath the lifted turf (you may need to stand in the pond to do this); or pile materials such as pebbles, logs, rocks or paving slabs on top of the liner edge. Concrete slabs have the added advantage of providing a 'hard standing' for a human viewing point.

Fill to the brim.

You now have a garden pond. Do not fret about the ugliness of the visible parts of the liner. A covering of a chemically inert substrate – washed gravel or children's play sand are ideal – will both hide the liner and provide footing for plants. Do not be tempted to use the soil from the hole as a base for plants since it will be too nutrient rich.

Time and algae will also cover over the liner.

There is no need to 'add' animals to the pond. They will find their own way, except for fish, and you really don't want these because they will eat all the other animals. All you need to do is plant vegetation in and around the pond.

Build a pond and the animals will come. During the Great War, pond life conquered the shell holes of the Western Front within a year, as *The Times* reported in 1917:

> But the strangest thing of all in nature's haste to hide the ravages of war seems to me the shell holes. As one wades through the deep herbage the lesser shell holes merely make the walking very difficult and uneven, for one's feet blunder among the shell holes, which are concealed by the growth, and trip over strands of barbed wire and unexploded shells and other things scattered everywhere out of sight.

Many of the larger holes, however, still remain half filled with water. Around the edges of the water white butterflies, which are thirsty creatures, crowd to drink, and when you disturb them they rise in clouds till the air is full of them, like a snowstorm. In the water itself a luxuriant pond life has developed. Little whirligig beetles dance mazy dances on the surface, and water boatmen swim about and water scorpions and other things just as in any village pond at home. I have spoken before of frogs in the new shell holes on the Vimy Ridge.

If pond animals can re-nature a shell hole of the Western Front they can re-nature a British peacetime pond.

Dragonflies are amongst the quickest colonizers, and are usually present in all medium-sized ponds with good water quality. A whole range of fauna will inadvertently be deposited by birds, including snails. Be aware that in transferring amphibians from another pond you may also be transferring pathogens.

I once dug a pond on our farm in the Black Mountains in October; by the following March it had newts, and by May swallows were skimming over the surface, open-mouthed for the midges, and the surface sparked with water beetles. At June twilight a Daubenton's bat sipped on the wing, and in September the heron paced the edge. Within a year the pond had an almost full set of pond life.

Stocking the Pond with Plants
A new pond will eventually become colonized by plants without human intervention. But few pond makers can bear to see their pond naked so a trip to the garden centre, a specialist water nursery, or an existing pond is almost certain. Always stock with native plant species, locally sourced. Most British pond plants are quite tough and can be introduced at any time of year. Some are rampant, thus you may wish to contain them in specialist plastic baskets. Or, better, terracotta pots.

A well-chosen selection of native flora is the crucial factor in making a pond a self-contained, self-regulating eco-system. Correctly chosen plants will keep the water oxygenated, de-algaed, without the need for costly electrical equipment. With the right plants you will have a habitat for an abundance of pond creatures. Almost all pond fauna need the sanctuary of vegetation, and the water plants themselves are an important source of food for everything from tadpoles to wildfowl.

Pond vegetation can be roughly divided into four

groups, according to the section of water edge that they occupy, and whether they are rooted or free floating.

Marginals: These are pondside plants that grow at the shallow edges of a pond, and often extend across land and water. Plant them in 2–5cm of water. They provide perches and food for hoverflies and bees, as well as safe zones for swimming things, including ducklings in their first few weeks after hatching. They also provide much food, especially seeds, for adult waterfowl. Try:

Marsh marigold (*Caltha palustris*): produces bright yellow flowers early in spring just as the first bees and other insects are emerging. Also known as kingcup. Spreads by creeping stalks and seeds, and is undemanding; moisture is all it requires. Grows up to 60cm in height, and its dense foliage provides necessary shadowy shelter for insects and amphibians. A must-have in a wetland wildlife garden.

Yellow flag (*Iris pseudacorus*).

Water forget-me-not (*Myosotis palustris*).

Watercress (*Rorippa nasturtium*): grows quickly, and has the added advantage that, if it needs to be hacked back, you can eat the cuttings. Wash well. Safest 'souped'.

Also: marsh pennywort, marsh cinquefoil, water dock, lesser spearwort.

Oxygenators: These are plants that live entirely or almost entirely underwater. Unfortunately most native oxygenators are intolerant of pollution, but the following submerged aquatic plants are all relatively robust. Oxygenators are important in the control of algae and duckweed.

Water milfoil (*Myriophyllium spicatum*): newts like to lay eggs under its spiked leaves. A good oxygenator.

Broad-leaved pondweed (*Potamogeton natans*): easy to establish, and much favoured by invertebrates.

Hornwort (*Ceratophyllum demersum*): jungly, beloved by all pond creatures.

Emergent plants: These have erect stems and leaves. As well as providing interesting natural 'architecture' they provide a ladder for dragonfly nymphs to crawl up preparatory to adulthood, as well as a place for adult dragonflies and damselflies to perch and roost. During the past forty years three species of dragonflies have become extinct in the UK. As they require good environmental conditions, dragonflies and damselflies are accurate indicators of a good pond habitat. Some flat stone or slate placed around the edge of your pond will enable dragonflies and damselflies to bask in the sunshine. Among the best emergent plants for the home pond:

Bulrush (*Typha latifolia*): to keep under control you may need to cut vigorously.

Bog bean (*Menyanthes trifoliata*): produces spikes of delicate pink flowers.

Greater pond-sedge.

Soft rush.

Branched bur-reed.

Floating plants: May be free-floating or some may have trailing roots, but in both cases their leaves float on the surface. The leaves provide beds for all manner of aquatic insects to mate on, as well as cooling shade for invertebrates in the water below. They will also help stop algae build-up. Consider:

White water lily (*Nymphaea alba*): to be favoured over the yellow water lily (*Nuphar lutea*), which has absolutely outsize leaves and an odd alcohol smell, hence 'brandy bottle' and 'brandy ball'. The exquisite linen-white flowers of *Nymphaea alba* are visited by countless species of butterfly and insect. You may need to control the lily by planting it in terracotta pots.

Amphibious bistort (*Persicaria amphibia*).

Bankside

Your pond will become even more of a wildlife wonder if you plant native flowers and trees adjacent to it, ideally in a two-metre-wide buffer strip. These will provide a foraging area for amphibians, and aquatic insects will seek out the flowers' nectar. Try devil's-bit scabious (*Succisa pratensis*), teasel (*Dipsacus fullonum*), hemp agrimony (*Eupatorium cannabinum*), and red valerian (*Centranthus ruber*). Cow parsley (*Anthriscus sylvestris*) provides excellent landing-pads for hoverflies.

Trees and shrubs provide shelter against adverse weather – our prevailing wind is from the south west but winter northerly winds are common – as well as nesting sites for pond birds. Additionally, small trees and bushes near the water's edge will provide perching sites for kingfishers and a steady supply of invertebrates to the pond, but as always with ponds take care not to over-shade. Useful pond trees? Alder (*Alnus glutinosa*) improves poor soils through nitrogen fixation, willow (*Salix spp.*) is easily propagated from cuttings, and bramble (*Rubus fruticosus*) is much favoured by moorhens as a home.

At the very least, let the grass near the pond grow

tall. This will make a welcome, if mundane-looking, habitat for many creatures, including the water vole (*Arvicola terrestris*), a priority species under the UK Biodiversity Action Plan, and Britain's fastest declining mammal. As well as suffering from loss of suitable habitat, the water vole – aka 'Ratty' in *The Wind in the Willows* – has been systematically predated by the invasive American mink since the latter's naturalization in Britain in the 1930s. In 1998 there were estimated to be only 875,000 individuals; during the Iron Age there were an estimated 6.7 billion water voles in Britain, and it was Britain's commonest mammal.

How to encourage Ratty to your pond? As well as the grassy buffer strip – which should be mown in autumn to prevent colonization by scrub – profile a section of the bank at 45 degrees, and ensure areas of the pond have a depth of water between 25 and 50cm.

Put a seat by the pond so you can comfortably enjoy watching all the dramas of pond life. Further afield, and within five hundred metres, a dry stone wall, a log pile, rock pile or hedge will provide a refuge for amphibians while they are travelling between sites, as well as a hibernaculum for others. Logs are really the five-star accommodation for newts because they create a damp micro-habitat with teeming supplies of insects for the amphibians to feast on.

A log pile will also accommodate toads, the organic gardener's friend. Toads, who are most active at night, feed on any moving prey they can swallow, generally invertebrates. This makes them useful in the garden when it comes to controlling populations of pest animals such as slugs and snails. Usually toads patiently sit and await their prey, which they catch with their long tongues, although they will roam if necessary.

When temperatures drop in the autumn, they will start to look for a suitable spot for hibernation. You can encourage toads to stay in the garden by providing a Toad Hall in the shape of said log pile, or alternatively a rock pile, or an old flower pot in the bottom of the hedge. Pack plenty of leaf litter round it.

Take care in autumn and winter before setting bonfires alight. Toads quite often hibernate in their lower depths.

Both adult and tadpole toads emit a toxin (bufo-toxin) from their skin to deter predators; it is known to cause irritation in humans. Toads should only be handled with gloves.

Plants to Avoid

Some invasive, non-native species will take over your pond. They can also escape into the wild and cause environmental mayhem by blocking waterways and appropriating the habitat of native plants. Their removal is expensive on the public purse; about £200 million is spent annually on removing non-native weeds. So, do not stock your pond with:

Floating pennywort (*Hydrocotyle ranunculoides*)

Parrot's feather (*Myriophyllum aquaticum*)

Water fern (*Azolla filiculoides*)

New Zealand pigmyweed/Australian swamp stonecrop (*Crasulla helmsii*)

Water hyacinth (*Eichhornia crassipes*)

Water lettuce (*Pistia stratiotes*)

Water primrose (*Ludwigia grandiflora*)

Make a Mini-Pond

Pretty much any waterproof container will make a pond. Wooden half-tubs and old stone sinks work well, or else choose a metal bucket or a glazed ceramic pot. Neither do ponds have to be 'dug in'. If your pond is free-standing just make sure there is a ramp in and out (stones, bricks and plants are ideal) so that wildlife has access. Newts and frogs will sometimes lay their eggs in a pond only a metre square, and if they can't be tempted to breed in your container pond it will at least help them keep cool in summer. Many other creatures will make use of it.

Fill your container with rainwater. If you have to use tap water, leave it to stand for a couple of days to neutralize.

Select your plants. As with a bigger pond, choose a combination of marginals, oxygenators, emergents and floating aquatics. Try water soldier (*Stratiotes aloides*) as an oxygenator, and dwarf lily (*Nymphaea odorata*) as a deep water aquatic. One lily will be enough for a container pond.

Plant any marginals in plastic aquatic baskets lined with gravel, filled with a specially designed compost (or clay loam), and dressed with gravel. To get the right height for the marginal plant use bricks. Place the planted basket on top.

Containers can also be used to spread and increase the wetland habitat around larger gardens. The more ponds the merrier.

Pond Management
The sort of pond you doubtless have in mind – clear water full of beetles, snails, newts, dragonflies, and a

bat circling overhead on a late summer's eve – requires some maintenance. Luckily, select the right plants and three quarters of the job is done for you. The other quarter you have to do.

Summer: People often panic if their pond starts drying out in the summer. Don't. Levels can dip dramatically without any negative results, and even drying up is unlikely to be a disaster. Annual drying out of ponds has been going on for millennia, and some species thrive on the process. Some of Britain's most interesting waterbodies are the temporary ponds of the New Forest, home to the fairy shrimp which is specially adapted for the drying-out process. A pond gone to desert will quickly re-colonize.

Indeed, since dragonflies lay their eggs around exposed pond margins it's best not to keep the pond topped to the brim in summer.

Overheating may be a problem in shallow ponds still resplendent with tadpoles. If the water temperature goes above 35°C tadpoles will die, so in this instance do add water, preferably rain water. Add little and often; a sudden flood of cold water will reduce the temperature of the pond, and consequently the body temperature of the amphibians in it. The shock can kill them.

Blanketweed and green algae tend to be a summertime problem in new ponds, dredged ponds and those positioned in direct sunlight or filled from the mains water supply. Remove blanketweed and algae by hand or with a rake. Algal blooms are usually short lived, and the pond should settle down as zooplankton increase and eat the algae. Persistent infestations can be controlled by reducing the level of nutrients in the water by removing bottom sediment in the winter (see below). Planting shade-casting flora species will also help. A quick fix can be had by putting a bale of barley straw into the pond. As the straw rots it releases chemicals that kill the algae; when the straw has turned black, about six months later, remove it. Unless the cause of the problem is addressed you will need to repeat the barley straw prescription.

Autumn: There are water weeds that multiply with unnerving ferocity. Be brutal with species bent on pond domination by pulling them up every year and reducing their presence by a good third. Do the job in autumn so as not to disturb breeding or hibernating animals. Aim always to keep at least 30 per cent of the pond surface open water.

Remember to leave the discarded vegetation by the pond's side for a day or two so that bugs and amphibians can escape back to the pond. After that, chuck it on the compost heap.

Over time, the inevitable build-up of fallen leaves and dead vegetation at the bottom of the pond may turn the water brown, as all the available oxygen is taken up by the process of decay. A layer of silt accumulates, which is no bad thing in itself as it is useful for hibernating frogs; eventually, however, the pond will become shallower and shallower ... until it becomes a muddy hollow, then scrub. With a small pond, you will need to dredge every five years or so; with a large pond every ten years. Again, this is a task for autumn and before amphibian hibernation begins. The silt from a small pond can be removed with a bucket, but a large pond may require a sludge pump. This can be hired from aquatic centres.

The dredged sludge may safely be added to flower borders.

To delay silting, small ponds can be covered with netting during the autumn to keep out falling leaves. Be sure to leave some of the surface free for wildlife access.

Winter: Since amphibians will hopefully be hibernating in the bottom of your pond, it must not freeze solid. (Ponds deeper than 60cm seldom suffer.) A frog in stasis can tolerate low oxygen levels; it cannot survive no oxygen levels. A tennis ball floating on top should prevent a great freeze-over. Never use boiling water to break the ice; positioning a saucepan of hot water on the surface until the ice has melted sufficiently is safer for pond life.

Spring: As soon as duckweed appears rake or pull it off by hand. New plants grow quickly in the warming water, so plant any friendly native species now. A planting rate of 1–3 per square metre will usually provide excellent cover in a year. Otherwise sit back and enjoy the amphibian mating and egg-laying season.

Wildfowl: If waterfowl visit your pond do not feed them white bread. It is junk food for ducks. Wholemeal bread is passable fodder, but better is grain or garden peas. Aside from lacking nutritional value, white bread if uneaten causes algal blooms, allows bacteria to breed and attracts rats.

Pond-Dipping
A net and a jam jar or shallow white oven dish are all that is needed for pond-dipping. Generations of children, not having a net handy, have made do by taking the foot and ankle from a pair of tights, using a

wire coat hanger to make the net's mouth, and sticking these into the end of a bamboo cane. Alternatively, use a kitchen sieve.

- Move your small-gauge net slowly through the pond.
- To identify your haul, tip contents either into a jam jar or, better still, a white dish. The white background is ideal to spot mini beasties against.
- Identify with a photographic guide.
- Don't forget to gently put back the animals as soon as possible. They can quickly become overheated or use up the oxygen in the water.
- Putting a drop of pond water under a microscope will open a whole universe.
- Try not to handle pond animals, not least because some of them bite.

Water Safety
Rather fewer children drown in ponds than our imagination would have us believe, but why take risks?

- Always supervise young children near a pond.
- Talk (calmly!) to children about the dangers of water; you want them to be aware not petrified.

- Wash hands after pond-dipping.
- A pond in a corner can be attractively fenced off to keep toddlers out whilst still letting most wildlife in. Think American-style picket fencing.
- If a pond is absolutely out of the question because of child safety, consider a bog garden.

A Pond Reading List

A. M. Abdualkader, A. M. Ghawi, M. Alaama, M. Awang, and A. Merzouk, 'Leech Therapeutic Applications', *Indian Journal of Pharmaceutical Sciences*, 75 (2), 2013

Frank Balfour-Browne, *Water Beetles and Other Things: Half a Century's Work*, 1963

BB (Denys Watkins-Pitchford), *Confessions of a Carp Fisher*, 1950

, *Brendon Chase*, 1944, reprinted 2000

, *Letters from Compton Deverell*, 1950

Trevor Beebee, *Pond Life*, 1992

John Clegg, *The Observer's Book of Pond Life*, 1956

O. G. S. Crawford, *Archaeology in the Field*, 1953

Christopher K. Currie, 'The Early History of the Carp and its Economic Significance in England', *The Agricultural History Review*, Vol. 39, no. 2, 1991

C. S. Elton, *The Pattern of Animal Communities*, 1966

Deryk Frazer, *Reptiles and Amphibians in Britain*, 1983

Ian Garrard and David Streeter, *The Wild Flowers of the British Isles*, 1983

Malcolm Greenhalgh & Denys Ovenden, *Freshwater Life* (Collins Pocket Guide), 2007

T. T. Macan, *A Guide to Freshwater Invertebrate Animals*, 1981

Edward Martin, *Dew-ponds: History, Observation, and Experiment*, 1914

L. C. Miall, *The Natural History of Aquatic Insects*, 1895

H. C. Prince, 'Pits and Ponds in Norfolk', *Erkunde*, 16, 1962

Oliver Rackham, *The History of the Countryside*, 1986

R. Maxwell Savage, *The Ecology and Life History of the Common Frog*, 1961

Carl Sayer et al, 'Managing Britain's Ponds – Conservation Lessons from a Norfolk Farm', *British Wildlife*, 25 (1), 2013

Peter Scott, *The Eye of the Wind*, 1961

Fred Slater, *The Common Toad*, 1992

Henry David Thoreau, *Walden*, 1854

J. Whitaker, *British Duck Decoys of Today*, 1918

Patrick J. Wisniewski, *Newts of the British Isles*, 1989

A Pond Playlist

Damon Albarn, 'Hollow Ponds', 2014

Henry Barter, 'Feeding the Ducks on the Pond', English traditional

Big Bill Broonzy, 'Crawdad's Song', American traditional

Frédéric Chopin, 'The Dragonfly', *Preludes*, Op. 28, no. 11, 1835–9

The Incredible String Band, 'Ducks on a Pond', 1968

Lemon Jelly, 'Nice Weather for Ducks', 2002

REM, 'Nightswimming', 1992

Camille Saint-Saëns, 'The Swan', *Carnival of the Animals*, 1886

Dmitri Shostakovich, 'The Dragonfly and the Ant', *Two Fables of Krylov*, Op. 4, no. 1, 1922

Jean Sibelius, 'The Swan of Tuonela', Op. 22, no. 2, 1895

Josef Strauss, 'The Dragonfly', Op. 204, 1866

Georg Philipp Telemann, *Violin Concerto in A Major* ('The Frogs'), TVW, 51:A4

Appendix – Contact Addresses

Government bodies (to be contacted for advice when managing a pond with protected species present)
ENGLAND Natural England, County Hall, Spetchley Road, Worcester WR5 2NP) Phone: 0300 0603900

SCOTLAND Scottish Natural Heritage, Great Glen House, Leachkin Road, Inverness IV3 8NW Phone: 01463 725000 Email: enquiries@nature.scot Website: www.nature.scot

NORTHERN IRELAND Northern Ireland Environment Agency, Klondyke Building, Cromac Avenue, Gasworks Business Park, Lower Ormeau Road, Belfast BT7 2JA Website: www.daera-ni.gov.uk

WALES Countryside Council for Wales Maes y Ffynnon, Penrhosgarnedd, Bangor, Gwynedd LL57 2DW Phone: 0845 1306 229 Website: http://natural resources.wales

Non-government bodies and charities

The British Association for Shooting and Conservation, Marford Mill, Rossett, Wrexham LL12 0HL Phone: 01244 573024 Email: conservation@basc.org.uk Website: www.basc.org.uk

British Dragonfly Society Email: conservation@british-dragonflies.org.uk Website: www.british-dragonflies.org.uk

Freshwater Habitats Trust, Bury Knowle House, North Place, Headington, Oxford OX3 9HY Phone: 01865 95505 Email: info@freshwaterhabitats.org.uk Website: www.freshwaterhabitats.org.uk The Freshwater Habitats Trust has begun to make networks of new ponds as part of the Million Ponds Project, with the ultimate aim of getting back to the million ponds that once enriched the British landscape a hundred years ago.

The Wildlife Trusts, The Kiln, Mather Road, Newark, United Kingdom NG24 1WT: Phone: 01636 677711 Email: enquiry@wildlifetrusts.org Website: www.wildlifetrusts.org

Acknowledgements

I had a pool – no, a pond – of helpers. So thank you: Tracy Pallant (especially), Julian Alexander, Susanna Wadeson, Sophie Christopher, Ella Horne, David Hill, Annie Robertson (merci!), Bella Bosworth, Bluebell, Tim Pugh, Kathryn and Nicholas Fox, Beci Kelly, Nick Hayes, Kate Samano, Josh Benn, Catriona Hillerton, Kate Tolley, Rachel Cross, all the Transworld sales team, Freda Lewis-Stempel, Tristram Lewis-Stempel, Penny Lewis-Stempel (of course).

John Lewis-Stempel is the only author to have won the Wainwright Prize for Nature Writing twice, for *Meadowland* and *Where Poppies Grow*. He was shortlisted a third time for *The Running Hare*, which was also shortlisted for the Independent Bookshop Week Adult Book Award and the Richard Jefferies Society Award. John writes a column on nature and farming for *Country Life* and was the 2016 BSME Magazine Columnist of the Year. He lives on the borders of England and Wales with his wife and two children.